Hans Andersen's Fairy Tales

Translated by
L. W. Kingsland

Illustrated by Rachel Birkett

Oxford University Press 1985   OXFORD   TORONTO   MELBOURNE

Oxford University Press, Walton Street, Oxford OX2 6DP
Oxford London
New York Toronto Melbourne Auckland
Kuala Lumpur Singapore Hong Kong Tokyo
Delhi Bombay Calcutta Madras Karachi
Nairobi Dar es Salaam Cape Town
and associated companies in
Beirut Berlin Ibadan Mexico City Nicosia
Oxford is a trade mark of Oxford University Press

© English Translation L. W. Kingsland 1985
First published 1985
ISBN 0 19 274532 8

British Library Cataloguing in Publication Data
Andersen, H. C.
Hans Andersen's fairy tales.
I. Title   II. Kingsland, L. W.
839.8'136[J]   PZ7
ISBN 0-19-274532-8

Phototypeset by Tradespools Limited, Frome, Somerset
Printed in Hong Kong

Contents

The Tinder-Box 1

Little Claus and Big Claus 10

The Princess and the Pea 24

Thumbelina 26

The Travelling Companion 39

The Little Mermaid 60

The Steadfast Tin-Soldier 85

The Emperor's New Clothes 91

The Wild Swans 97

The Ugly Duckling 115

The Nightingale 126

The Swineherd 137

The Fir-Tree 143

The Snow Queen 153

The Red Shoes 188

The Shepherdess and the Chimney-Sweep 195

The High Jumpers 201

The Little Match-Girl 204

The Happy Family 208

Everything in its Right Place 213

What the Old Man Does is Always Right 222

The Goblin at the Grocer's 228

Five From a Pea Pod 233

Soup on a Sausage Stick 237

Absolutely True! 252

The Silver Shilling 255

Colour Plates

facing page

6   Whew! There sat the dog with eyes as big as millstones!

22   "So it's you, Little Claus? First of all, here are some cattle for you".

54   When they had danced for some time the princess told the troll she had a new suitor.

70   There she would sit and look at the young prince who thought he was quite alone in the clear moonlight.

86   Now the paper parted and the tin soldier fell through—and he was at once swallowed by a big fish.

102   She thought she was still dreaming, so strange did it seem to her to be carried over the sea, high up through the sky.

118   The duckling thought they meant to hurt him and in his fright he flew right into the milk-pan.

134   He opened his eyes and then he saw that it was Death sitting there wearing his golden crown and holding the Emperor's golden sword in one hand and his splendid standard in the other.

166   He spent his time dragging sharp flat pieces of ice about, arranging them in all sorts of ways, and trying to make something of them.

198 The sky with all its stars was over their heads and below them all the roofs of the town.

230 A bright beam of light rose out of the book and grew into a trunk, a mighty tree that rose up high and spread out its broad branches over the student.

246 'Fantasy came, the feather was plucked out and I seized it,' said the little mouse.

Dedication

This second and revised edition is dedicated to the Oxford University Press who first encouraged me some twenty-five years ago to translate Andersen's tales, and to their editors for their unfailing patience and courtesy.

Hans Andersen wrote over one hundred and sixty Fairy Tales and Stories, and yet the task of varying the content of a selection such as this is a particularly difficult one. Andersen's reputation outside Denmark deservedly rests upon a dozen or so well-known tales which every reader will expect to find and which must therefore hold their place, and these are undoubtedly his best. There are interesting and sometimes powerful stories in Andersen that might still make good reading for adults, and there are those where to read them sympathetically one has to accept the sentimentality and morbid piety of his age. There is a fair number left from which to choose, but their conflicting attractions and weaknesses make that choice difficult. It would be easier to produce a volume twice the size of this one.

I have endeavoured to reproduce the simplicity and directness of Andersen's Danish in good idiomatic English that can easily be read aloud.

The Tinder-Box

A SOLDIER came marching down the road—left, right! left, right! He had his pack upon his back and a sword by his side, for he had been to the wars and was now on his way home. On the road he met an old witch; she was so ugly her bottom lip hung right down upon her breast.

'Good evening, soldier!' she said. 'That's a fine sword you've got there, and a great pack, too—you look a proper soldier, if ever there was one! And now, if you do as I say, you shall have as much money as you like.'

'Thanks very much, old witch,' said the soldier.

'Do you see that big tree?' said the witch, as she pointed to the tree that was standing by their side.

'It's quite hollow inside. Climb up to the top of the trunk and you'll see a hole you can slip through. Then you can drop right down into the tree. I'll tie a cord round your waist so that I can hoist you up again when you shout for me.'

'What have I got to do when I'm down in the tree?' asked the soldier.

'Fetch the money!' said the witch. 'When you get down to the bottom of the tree, you'll find yourself in a large passage. It's quite light, for there are more than a hundred lamps burning there. Then you'll see three doors: you can open them because the keys are in the locks. Go into the first room and you'll see in the middle of the floor a big chest with a dog sitting on it. He's got a pair of eyes as big as tea-cups—but don't let that worry you. I'll give you my blue-checked apron; you can spread it out on the floor, go boldly up to the dog, take hold of him and set him down on my apron. Then open the chest and take out as much money as you like. It's all coppers; but if you'd rather have silver, you'll have to go into the next room. The dog that sits there has a pair of eyes each as big as a millstone, but don't let that worry you—put him on my apron and help yourself to the money. If you want gold, however, you can get that, too—and as much as you can carry—when you go into the third room. But the dog that sits on the money-chest there has two eyes each as big as the Round Tower in Copenhagen. He's a real dog, you can take it from me! But don't you worry about that. Just put him down on my apron and he'll do you no harm. And then you can take as much gold as you like from the chest.'

'That doesn't sound so bad!' said the soldier. 'But what have I got to give you, old witch? I'm pretty certain you want something out of it!'

'No,' said the witch, 'not a single penny do I want. All you have to do is to fetch me an old tinder-box my grandmother left behind the last time she was down there.'

'Well, let me get the cord round my waist!' said the soldier.

2

'Here it is,' said the witch. 'And here's my blue-checked apron!'

So the soldier climbed up the tree and dropped down the hole. There he stood, just as the witch had told him, down below in a large passage where hundreds of lamps were burning.

Then he unlocked the first door. Ooh! There sat the dog with eyes as big as tea-cups glaring at him.

'Nice dog. Good boy,' said the soldier. He put him on the witch's apron and took as many coppers as he could carry in his pockets: then he shut the chest, put the dog back again and went into the second room. Whew! There sat the dog with eyes as big as millstones!

'You shouldn't look at me so hard,' said the soldier. 'You might get eyestrain.' And so he put the dog on the witch's apron, but when he saw the mass of silver coins in the chest, he threw away all the coppers he had taken, and filled his pockets and his pack with nothing but silver. Then he went into the third room. Oh, it was terrible! The dog there really did have two eyes each as big as the Round Tower—and they were turning round in his head just like wheels!

'Good evening,' said the soldier and touched his cap, for he had never seen a dog like that before. But when he had looked at him a bit, he thought he had better get a move on, so he lifted him on to the floor and opened the chest. Heavens above, what masses of gold there were! With that he could buy the whole of Copenhagen, and all the sugar-pigs, tin-soldiers, whips, and rocking-horses in the world as well. That was something like money! And now the soldier threw away all the silver coins he had filled his pockets and his pack with, and took the gold instead—yes, he filled all his pockets, his pack, his cap, and his boots so that he could hardly walk. He had money enough now. He put the dog back on the chest, slammed the door to, and shouted up the tree,

'Haul me up now, old witch!'

'Have you got the tinder-box?' asked the witch.

''Strewth!' said the soldier. 'I've clean forgotten it!' Then he went and fetched it. The witch hauled him up, and so he stood upon the road once more with pockets, boots, pack, and cap all crammed full of money.

'What do you want with that tinder-box?' asked the soldier.

'That's nothing to do with you!' said the witch. 'You've got your money—now give me my tinder-box.'

'Rubbish!' said the soldier. 'Either tell me right now what you want with it, or I'll draw my sword and hack your head off.'

'No, I won't!' said the witch.

So the soldier cut her head off. And there she lay. Then he tied all his money up in her apron, heaved the bundle on to his back, put the tinder-box in his pocket, and marched straight off to the town.

It was a fine town, and he took himself into the finest inn, where he booked the very best rooms, and ordered all the things he liked best to eat, for he was rich now with all that money.

The boot-boy who took his boots to clean thought it decidedly odd that such a rich gentleman should have such an old pair. He had not yet bought himself new ones, but the next day he got new boots and fine new clothes to wear.

The soldier had now become a gentleman of note, and people told him about all the fine things to be found in their town, and about their king, and what a lovely princess his daughter was.

'How can you get to see her?' asked the soldier.

'It's quite impossible to get to see her,' everybody said. 'She lives in a great castle made of copper, with ever so many walls and towers round it. No one but the king dare go in and out to her, because it's been foretold that she'll marry an ordinary common soldier, and the king doesn't like the idea at all.'

4

'I'd very much like to see her,' thought the soldier, but that was something he would certainly never get leave to do.

He was living a life of pleasure now, going to plays, riding in the royal gardens, and giving a great deal of money to the poor—and that was noble of him. He knew well enough from the old days how wretched it can be not to have a penny. He was rich now, he had fine clothes, and he made very many friends who all said he was a rare fellow, a proper gentleman—and that pleased the soldier greatly. But since he was spending his money every day and not getting a penny to replace it, he was left at last with no more than two shillings.

He had to move from the beautiful rooms where he had been living, up to a narrow little attic right under the roof. He had to brush his own boots and mend them with a darning-needle, and none of his friends would come to visit him because there were far too many stairs to go up.

It was quite dark in the evening, and he could not even buy himsef a candle, but then he remembered that there was a little stump in the tinder-box he had taken from the hollow tree the witch had helped him to go down. He got the tinder-box and took out the candle-end, but just as he was striking a light and the sparks were flying from the flint the door sprang open, and the dog with eyes as big as tea-cups, which he had seen down under the tree, stood before him.

'What are my master's commands?' asked the dog.

'What on earth!' said the soldier. 'This tinder-box is certainly something, if I can get what I want like this. Fetch me some money,' he said to the dog. He was off in a flash. In another flash he was back again, holding a big bag full of coppers in his mouth.

The soldier now knew what a delightful tinder-box it was. If he struck once, the dog that sat on the chest of copper coins came; if he struck twice, the one that had the silver coins came; and if he struck three times, the one with the gold came. And now the soldier moved down into his beautiful rooms again and went about in good clothes, and so all his friends recognized him immediately, and they were all very fond of him once more....

Then one day he thought, 'It's a very odd thing, though, that you can't get to see the princess. They all say how lovely she must be. But what's the good of that when she's always shut up in that great copper castle with all those towers. Can't I get to see her somehow? ... I've got it!—Where's my tinder-box!'

And so he struck a light, and in a flash there was the dog with eyes as big as tea-cups.

'I know it's the middle of the night,' said the soldier, 'but I would so love to see the princess, just for a second.'

The dog was outside the door at once, and before the soldier had time to realize it, he saw the dog was back again with the princess. She was sitting sound asleep on the dog's back, and she was so lovely anyone could see she was a real princess. The soldier could not stop himself from kissing her, for he was a real soldier.

Then the dog ran back again with the princess, but when the morning came and the king and queen were pouring out tea, the princess said she had had such a strange dream during the night about a dog and a soldier. She had ridden on the dog, and the soldier had kissed her.

'That's a fine story, I'm sure!' said the queen.

One of the old ladies-in-waiting now had to keep watch by the princess's bed the next night to see whether it had really been a dream or what else it could be.

The soldier longed terribly to see the lovely princess again, and so the dog came during the night, took her and ran off as fast as he could. But the old lady-in-waiting put her wellingtons on, and ran just as hard after him. When she saw them disappear into a large house, she thought, 'Now I know where it is!' And she drew a large cross on the gate with a piece of chalk. Then she went home and lay down, and the dog came back, too, with the princess. But when he saw that a cross had been drawn on the gate where the soldier

lived, he took a piece of chalk, too, and marked crosses on all the gates throughout the whole town. And that was a clever thing to do, for now the lady-in-waiting would certainly not be able to find the right gate when there were crosses on all of them.

Early in the morning the king and the queen, the old lady-in-waiting, and all the officers of the court went to see where it was the princess had been to.

'There it is!' said the king when he saw the first gate with a cross on it.

'No, my dear husband, it's there!' said the queen, who was looking at the second gate with a cross on.

'But there's one there—and one there!' they all cried together, and wherever they looked, there were crosses on the gates. Then they could see it was not any use trying to find it.

Now the queen was a very clever woman, who could do other things besides riding in a coach. She took her big pair of gold scissors, cut out some pieces of silk, and made a pretty little bag from them. She filled it with fine small grains of buckwheat, tied it on the princess's back, and when that was done, she clipped a little hole in the bag so that the grains would be sprinkled all the way along where the princess went.

During the night the dog came again, took the princess on his back, and ran off with her to the soldier, who was so very fond of her and wished so much he had been a prince so that he could have her for his wife.

The dog did not notice the grain sprinkled all along the road from the castle right up to the soldier's window, where he leapt up the wall with the princess. But in the morning the king and queen saw clearly where their daughter had been, and so they seized the soldier and threw him into prison.

And there he sat. Oh, how dark and miserable it was— and added to that, they said to him, 'Tomorrow you must be hanged!' That was not at all funny—and he had left his tinder-box behind at the inn. In the morning he looked between the iron bars of his little window and could see people hurrying out of town to see him hanged. He heard the drums and saw the soldiers marching by. Everybody was off to watch, and among them was a cobbler's boy wearing his leather apron and a pair of slippers. He was trotting along at such a gallop that one of his slippers flew off and landed against the wall where the soldier sat peering out between the iron bars.

'Hi, you cobbler's boy! There's no need to be in such a hurry,' the soldier said to him. 'There won't be anything doing before I get there. Just run along to where I live and fetch me my tinder-box, and I'll give you a shilling. But you must make good use of your legs.'

The cobbler's boy was anxious to have his shilling and scurried off after the tinder-box. He gave it to the soldier—and now we shall hear what happened!

Outside the town a great gallows had been built, and all round it stood the soldiers and hundreds of thousands of people. The king and queen were sitting on a beautiful throne immediately opposite the judge and the whole council.

The soldier had already mounted the ladder, but just as they were going to fasten the rope about his neck, he said that a wrong-doer was always granted a harmless request before he underwent his punishment. He would dearly love to smoke a pipe of tobacco— it would be his last pipe in this world.

Now the king could not say no to that, and so the soldier took his tinder-box and struck a light—one, two, three! And there stood all the dogs, the one with eyes as big as tea-cups, the one with eyes like millstones, and the one that had eyes as big as the Round Tower!

'Now help me, so that I shan't be hanged!' said the soldier, and so the dogs pounced upon the judge and the whole council, took one by the legs and one by the nose, and hurled them many fathoms up in the air so that when they fell down they were smashed to pieces.

'I will not ...!' said the king, but the biggest of the dogs seized both him and the queen, and hurled them up after the rest of them. The soldiers were terrified, and all the people cried out, 'Little soldier you shall be our king and have our lovely princess!'

So they sat the soldier in the king's coach, and the three dogs all frisked about in front of him and shouted, 'Hurray!' And the boys whistled through their fingers, and the soldiers presented arms. The princess left the copper castle and became queen—she liked that! The wedding lasted a week, and the dogs sat with them at table, their eyes wide with astonishment.

Little Claus and Big Claus

THERE lived in a village two men with the same name. Both were called Claus, but one of them owned four horses, and the other only one. Now to distinguish them one from the other, the one who had four horses was called Big Claus, and the one who had only one was called Little Claus. And now we shall hear what happened to them, for this is a true story.

The whole week through, Little Claus had to plough for Big Claus and lend him his one horse; in return Big Claus helped him with all four of his, but only once a week, and that was on Sundays. Hurray!— How Little Claus cracked his whip over all five horses, for they were certainly as good as his for that one day. The sun was shining beautifully, and all the bells in the church-tower were ringing for church. As the people of the village, all in their Sunday best, went by with their hymn-books under their arms, on their way to hear the parson preach his sermon, they looked at Little Claus ploughing there with five horses, and he was so delighted he cracked his whip again and again, and shouted, 'Gee-up, there, all my horses!'

'You mustn't say that!' said Big Claus. 'You know quite well only one of them is yours.'

But as soon as someone else went by on his way to church, Little Claus forgot that he mustn't say it, and shouted again, 'Gee-up, there, all my horses!'

'Now,' said Big Claus, 'I must ask you to stop it! If you say that once again, I shall strike your horse over the head. Then he'll fall dead on the spot, and that'll be the end of him!'

'I really won't say it any more,' said Little Claus, but as soon as people went past again and nodded good day to him, he was so delighted and thought it looked so grand to have five horses to plough his field with, that he cracked his whip once more and shouted, 'Gee-up, there, all my horses!'

'I'll gee your horses up!' said Big Claus, and he took a mallet and struck Little Claus's horse over the forehead so that it fell down quite dead.

'Oh, now I've no horse at all!' said Little Claus, and started crying. Shortly afterwards he flayed his horse, took the hide and let it dry thoroughly in the wind. Then he put it in a bag, slung it over his shoulders and trudged off to the town to sell his horsehide.

He had a very long way to go. He had to go through a great dark wood, and the weather had now turned bad. He lost his way completely, and before he came upon the right road again it was evening, and it was much too far either to go on to the town or to turn back home before it was night.

Close by the roadside there lay a large farmhouse. The shutters outside the windows were closed, but a light was shining out over the top of them. 'Perhaps they might let me stay the night here,' thought Little Claus. So he went up to the door and knocked.

The farmer's wife opened the door, but when she heard what he wanted, she said he'd better be on his way, her husband was not at home, and she didn't let strangers in.

'Well then, I must find somewhere to lie down outside,' said Little Claus, and the farmer's wife shut the door in his face.

Close by stood a haystack, and between it and the house was built a little shed with a flat thatched roof.

'I can lie down up there,' said Little Claus, as he noticed the roof. 'That'll make a very nice bed, so long as the stork doesn't fly down and nip me in the leg.' For there was a stork standing wide awake up on the roof, where it had its nest.

Then Little Claus climbed up on to the shed, and there he lay twisting and turning until he was comfortable. The wooden shutters in front of the windows did not meet properly at the top, and so he could see right into the living-room.

There was a great table set there, with wine and a roast joint and a lovely fish. The farmer's wife and the parish clerk were sitting at the table, and there was no one else there at all. She was pouring out for him, and he was tucking into the fish, which was something he was very fond of.

'I'd like a bit of that, too!' said Little Claus, craning his neck towards the window. What lovely cakes he could see down there on the table! It was a proper feast.

Then he heard someone riding along the road towards the house—it was the farmer coming home.

The farmer was a real good-hearted man, but he had one odd weakness—he just couldn't bear to see a parish clerk. If ever he caught sight of one, he became quite frantic. And so that was why the parish clerk had come to say good evening to the farmer's wife while he knew her husband was away from home, and the good woman had set before him all the nicest things she had to eat.

When they heard her husband coming, they were so terrified that the wife told the parish clerk to crawl into a big empty chest that stood over in the corner. And that's what he did, for he was well aware that the poor farmer couldn't stand the sight of a parish clerk. The farmer's wife hastily hid all the lovely food and wine inside her baking-oven, for if her husband caught sight of them, he would undoubtedly ask what it was all about.

'Ah, well!' sighed Little Claus, up on the shed, as he saw all the food put away.

'Is there anyone up there?' asked the farmer, peering up at Little Claus. 'What are you lying up there for? You'd better come in with me.'

So Little Claus told him how he had lost his way, and asked if he could put him up for the night.

'Why, of course!' said the farmer. 'But we must have a bite to eat first.'

The farmer's wife seemed very pleased to see them both, spread a cloth on a long table, and gave them a large basin of porridge. The farmer was hungry and ate with a good appetite, but Little Claus couldn't stop thinking of the lovely roast joint, fish, and cakes he knew were standing inside the oven.

He had put the sack with the horsehide down by his feet under the table, for, you remember, it was to sell this in the town that he had left home that day. He didn't fancy the porridge a bit, and so he trod on the sack and the dry skin inside squeaked quite loudly.

'Sh!' said Little Claus to his sack, but at the same time he trod on it again, and it squeaked much louder than before.

'What's that you've got in your sack?' asked the farmer.

'Oh, that's a wizard,' said Little Claus. 'He says there's no need for us to eat porridge. He's conjured up a whole ovenful of roast beef, fish, and cake.'

'What!' cried the farmer, and had the oven open in a

13

flash. There he saw all the good food his wife had hidden, but of course he believed the wizard in the sack had conjured it up for them. His wife dared not say a word. She set the food on the table at once, and they ate the fish and the roast beef and the cake. Little Claus thereupon trod on the sack again and made the hide squeak.

'What does he say now?' asked the farmer.

'He says,' said Little Claus, 'that he's also conjured up three bottles of wine for us, and they're standing in the oven, too.' And now the farmer's wife had to take out the wine she had hidden, and the farmer drank and grew quite merry. He would, he thought, dearly love to own a wizard like the one Little Claus had in his sack.

'Can he conjure up the devil, too?' asked the farmer. 'I'd like to see him, for I'm feeling merry now.'

'Yes,' said Little Claus, 'my wizard can do anything I ask him. True, isn't it?' he asked, treading on the sack and making it squeak. 'Did you hear him say, "Of course!"? But the devil looks horrible. I shouldn't ask to see him if I were you.'

'Oh, I'm not frightened of him. What does he look like?'

'Why, he'll be the spitting image of a parish clerk!'

'Ugh!' said the farmer. 'That'd be horrible! I'd better warn you that I just can't stand the sight of a parish clerk. But now that I know that it's only the devil, I shan't mind so much! I've got my courage back now—but he mustn't come too near me.'

'I'll ask my wizard now,' said Little Claus, treading on his sack and bending his ear down to listen.

'What's he say?'

'He says if you go and open that chest over there in the corner, you'll see the devil skulking inside, but you must keep hold of the lid so that he can't slip out.'

'Will you help me to hold it?' asked the farmer, as he went over to the chest where his wife had hidden the real parish clerk, who was crouching inside, much afraid.

The farmer lifted up the lid a little way and peeped under it. 'Ugh!' he shrieked, springing backwards. 'You're right! As I caught sight of him, he looked just like our parish clerk. Oh, it was ghastly!'

They had to have a drink to recover, and then they went on drinking far into the night.

'You must sell me your wizard,' said the farmer. 'Ask whatever you like for him. Tell you what, I'll give you a bushel of money right away.'

'No, I couldn't do that,' said Little Claus. 'Just think how much my wizard can do for me!'

'Ah, I'd dearly love to have him,' said the farmer. And he went on pleading.

'Well,' said Little Claus at last, 'since you've been good enough to give me house-room for the night, I don't mind if I do. You shall have the wizard for a bushel of money—but the bushel must be brim-full, mind.'

'That it shall,' said the farmer. 'But you must take the chest away with you. I'll not have it in the house an hour longer—you can never tell but what he's still inside it!'

Little Claus gave the farmer the sack with the dry hide inside it, and got a bushelful of money, top measure, too, in exchange. The farmer also made him a present of a large wheelbarrow to carry away the money and the chest.

'Good-bye!' said Little Claus, as he wheeled off the money and the chest where the parish clerk was still crouched.

On the other side of the wood was a deep wide stream, its current running so strongly that one could hardly swim against it. A big new bridge had been built over it, and Little Claus stopped in the middle of it and said quite loudly so that the parish clerk inside the chest could hear him:

'What on earth do I want with this damn-fool chest? It's so heavy, it might be full of stones! I'm much too tired to push it any further, so I'll just tip it over into the stream. If it floats down to the house, well and good; if it doesn't, it's all the same to me.'

Then he took hold of the chest with one hand and lifted it a little, as if he were about to push it over into the water.

'No! No! Stop it!' cried the parish clerk from inside the chest. 'Let me out first!'

'Ooh,' said Little Claus, pretending to be frightened, 'he's still there! I must topple it over into the stream right away and drown him.'

'Oh, no, no!' cried the parish clerk. 'I'll give you a whole bushelful of money if you'll only stop!'

'Well, that's another matter,' said Little Claus, and opened the chest. The parish clerk crawled out at once, kicked the empty chest into the water, and went to his house, where he gave Little Claus a whole bushelful of money. With the one he had already got from the farmer, he now had a whole wheelbarrow full!

'Well, I've been pretty well paid for my horse,' he said to himself, when he was home again in his own room, and had tipped out all the money in a great heap in the middle of the floor. 'Won't Big Claus be annoyed when he gets to know how rich I've become from my one horse—but I won't tell him right out just like that.'

Then he sent a boy over to Big Claus to borrow a bushel-measure.

'I wonder what he wants with that,' thought Big Claus, and smeared some tar on the bottom so that a little of whatever was measured in it should stick to it. And it did, too, for when he got his bushel-measure back, there were three new half-crowns sticking to it.

'What on earth!' said Big Claus, and ran straight over to Little Claus. 'Where have you got all that money from?'

'Oh, I got that for my horsehide I sold last night.'

'You got a pretty good price for it, I'll say!' said Big Claus. He ran home, took an axe and struck all four of his horses over the head with it. Then he flayed them and drove into the town with the hides.

'Hides! Hides! Who'll buy my hides!' he cried through the streets.

All the cobblers and tanners came running out and asked what he wanted for them.

'A bushel of money each,' said Big Claus.

'Are you mad?' they all said. 'Do you think we have money by the bushel?'

'Hides! Hides! Who'll buy my hides!' he cried again, but to everyone who asked what the hides cost he answered, 'A bushed of money.'

'He's trying to make fools of us,' they all said, and so the cobblers took their straps and the tanners their leather aprons, and they all set about Big Claus.

'Hides! Hides!' they mocked. 'We'll give you a hide all right—one that will spit red! Out of the town with him!' they cried, and Big Claus, thrashed as he had never been before, had to take to his sheels for all he was worth.

'All right!' he said when he got home. 'Little Claus shall pay for this. I'll strike him dead!'

But Little Claus's grandmother lay dead in the house. True, she had always been very ill-tempered and unkind to him, but he was still quite sorry. He took the dead woman and laid her in his own warm bed to see if that would bring her back to life again. He would let her stay there all night, while he himself would sit on a chair over in the corner and go to sleep, as he had often done before.

During the night while he was sitting there, the door opened and in came Big Claus with his axe. He knew quite well where Little Claus's bed stood, and so he went straight over to it and struck the dead grandmother over the head, believing, of course, that it was Little Claus.

'There!' he said. 'Now you won't make a fool of me again.' And so he went back home.

'He's a bad wicked man,' said Little Claus. 'He wanted to strike me dead. It's a good thing the old woman was dead already, otherwise he'd have killed her.'

Then he dressed his old grandmother up in her Sunday best, borrowed a horse from one of the neighbours, harnessed it to the cart, and sat his old grandmother up in the back seat so that she couldn't fall out when he was driving, and then they drove off through the wood. When the sun rose, they were outside a large inn, where Little Claus drew up and went in to get something to eat.

The innkeeper was very, very rich. He was a very good man, too, but so quick-tempered you would have thought he was made of pepper and snuff.

'Good morning!' he said to Little Claus. 'You're up very early today in all your best clothes.'

'Yes,' said Little Claus, 'I have to go into town with my old grandmother. She's sitting outside in the cart. I can't get her inside, but would you mind taking her a glass of mead? You must speak fairly loud to her. Her hearing's none too good.'

'Righto, I will,' said the innkeeper, and he poured a large glass of mead which he took out to the dead grandmother propped up in the cart.

'Here's a glass of mead from your grandson,' said the inkeeper, but the dead woman said never a word and sat quite still.

'Can't you hear?' bawled the innkeeper as loud as he could. 'Here's a glass of mead from your grandson!'

Once more he shouted the same thing, and then yet again. But as she did not make the slightest movement, he lost his temper and threw the glass right in her face so that the mead ran down her nose and she fell backwards into the cart, for she was only propped up and not fastened in any way.

'Good God!' cried Little Claus, rushing out of the door and seizing the innkeeper by his lapels. 'You've killed my grandmother! Just look at that great hole in her forehead!'

'Oh dear, oh dear, it was an accident!' wailed the innkeeper, beating his hands together. 'It's all because of my bad temper! My dear Little Claus, I'll give you a whole bushel of money and have your grandmother buried as if she were my own, if only you'll say nothing—otherwise they'll have my head off, and that'd be frightful!'

So Little Claus got a whole bushel of money, and the innkeeper buried his old grandmother as if she were his own.

Now when Little Claus got home again with all this money, he sent his lad over to Big Claus right away to ask if he might borrow a bushel-measure.

'What on earth?' said Big Claus. 'Haven't I killed him after all? I must look into this myself.' And so he took the measure over to Little Claus.

'My, where did you get all this money from?' he asked, opening his eyes wide at the sight of all the money.

'It was my grandmother you struck dead—not me,' said Little Claus. 'I've just sold her and got a bushel of money for her!'

'That was a good price, I must say,' said Big Claus, and he hurried home, took an axe and struck his own old grand-mother dead straight away. He sat her up in a cart, drove into town to where the apothecary lived and asked if he would like to buy a dead body.

'Who is it, and where have you got it from?' asked the apothecary.

'It's my grandmother,' said Big Claus. 'I hit her over the head so that I can get a bushel of money for her.'

'Heaven preserve us!' said the apothecary. 'You can't know what you're saying! For God's sake, don't say things like that or you'll lose your head.' And then he told him plainly what a terribly wicked thing he had done, and what a bad man he was, and how he ought to be punished. Whereupon Big Claus was so terrified, he rushed straight out of the apothecary's shop and into his cart, whipped up the horses and made for home. But the apothecary and every-body else thought he was mad, and so they let him go wherever he would.

19

'You shall pay for this!' said Big Claus, when he was out on the open road. 'Yes, you shall pay for this all right, Little Claus!' And then, as soon as he got home, he took the biggest sack he could find, went over to Little Claus and said, 'Now you've made a fool of me again! First I kill my horses, and then I kill my old grandmother. And it's all your fault! But you shall never make a fool of me again.' And then he took hold of Little Claus round the waist and thrust him into the sack. He heaved him up on his shoulders and cried, 'Now I'm going out to drown you!'

It was quite a long way to go before he came to the stream, and Little Claus was not so light to carry. The road went close by the church, where the organ was playing and the congregation singing. So Big Claus put down the sack with Little Claus inside close by the church door. He thought it would be rather nice to go in and listen to a hymn first before he went any farther. Little Claus couldn't possibly get out, and everyone was in church. So in he went.

'Oh, dear! Oh, dear!' sighed Little Claus inside the sack. He turned and he twisted, but it was quite impossible to loosen the string. At that moment an old cattle drover came along. He had snow-white hair and he leant upon a stout staff. He was driving a whole drove of cows and bulls in front of him, and some of them bumped against the sack where Little Claus sat huddled up, and knocked it over.

'Oh, dear!' sighed Little Claus. 'I'm so young, and I'm on my way to heaven already.'

'And poor old me!' said the cattle-drover. 'I'm so old, and I can't get there yet.'

'Open the sack!' cried Little Claus. 'If you crawl inside and take my place, you'll get to heaven right away.'

'I'd dearly love to!' said the cattle-drover, and opened the sack for Little Claus who scrambled out at once.

'Will you look after the cattle?' said the old man, as he crawled into the sack. Little Claus tied it up, and then went on his way with all the cows and bulls.

A little later Big Claus came out of the church. He heaved the sack upon his shoulders again, and thought it had become much lighter. He was quite right, for the old cattle-drover was scarcely half the weight of Little Claus. 'He feels a lot lighter now—it must be because I listened to that hymn.' So he went on to the stream that ran deep and wide, threw the sack with the old cattle-drover inside over into the water, and shouted after him—he still believed he was Little Claus, of course—'There, now you shan't make a fool of me any more.'

So he turned homewards, but when he came to the cross-roads, he met Little Claus driving his cattle along.

'What on earth!' cried Big Claus. 'Haven't I just drowned you?'

'Yes, of course you have,' said Little Claus. 'You threw me into the river not half an hour ago.'

'But where did you get all those fine cattle from?' asked Big Claus.

'They're sea-cattle,' said Little Claus. 'I must tell you the whole story—but first I must thank you for drowning me, for I'm well off now. I'm really rich, believe me.

'I was very frightened when I was in that sack, and the wind whistled in my ears as you threw me over the bridge into the cold water. I sank at once to the bottom, but I didn't hurt myself, for it was covered in fine soft grass. I landed on this, and the sack was opened at once, and the loveliest young lady in a snow-white dress with a green wreath on her wet hair took me by the hand and said, "So it's you, Little Claus? First of all, here are some cattle for you. And five or six miles along the road, there's another drove waiting—and I'm going to make you a present of them, too." Then I saw that the river was like a main road for the sea-folk. Down there on the river-bed they were walking and riding all the way from the sea right up into the land where the stream comes to an end. It was lovely down there, with flowers and fresh grass and fishes swimming in the water and playing

21

round my ears, just like birds in the sky. They were such nice-looking people there, and you should have seen the cattle walking along by the ditches and fences!'

'But why have you come up here again so soon?' asked Big Claus. 'I shouldn't have, if it was so nice down there.'

'Well,' said Little Claus, 'that's just my artfulness! You remember I told you the sea-lady said that five or six miles up the road—and by road she meant the river, of course, because she can't go anywhere else—there was another whole drove of cattle awaiting me? But you know how the river twists and turns, now this way, now that, taking you a long way out of your way. If you can do it, it's much shorter to cut across the land and join the river again higher up. I shall save about half the journey that way, and get to my sea-cattle all the more quickly.'

'What a lucky fellow you are!' said Big Claus. 'Do you think I should get some sea-cattle if I went down to the bed of the river?'

'Yes, I should think so,' said Little Claus. 'But I can't carry you in a sack all the way to the river. You're too heavy for me. But if you like to walk there by yourself and then crawl into the sack, I'll throw you into the water with the greatest pleasure.'

'Thank you very much,' said Big Claus. 'But, mind, if I don't get any sea-cattle when I get down there, I'll give you a good hiding, you trust me!'

'Oh, please, you wouldn't be so unkind!' And so they walked along to the river. As soon as the cattle, who were thirsty, saw the water, they ran as hard as they could so that they could go down and drink.

'Look at the hurry they're in,' said Little Claus. 'They're longing to get down on the river-bed again.'

'Yes, but you must help me first!' said Big Claus. 'Or else you'll get that hiding!' And so he crawled into a big sack which they had found lying across the back of one of the

bulls. 'Put a stone in, else I'm afraid I shan't sink,' said Big Claus.

'You'll do that all right!' said Little Claus, but he put a big stone in the sack all the same, tied the string tight, and then gave it a good push. Splash! Over went Big Claus into the river and he sank down to the bottom at once.

'I'm afraid he won't find his sea-cattle,' said Little Claus, as he made his way home with what he had got.

The Princess and the Pea

THERE was once a prince who wanted to marry a
princess. But she had to be a real princess. So he
travelled all round the world to find one, but there
was always something wrong—there were plenty of
princesses, though whether they were real princesses he could never quite find out: there was always
something that was not just right. So he came
home again and was very unhappy because he
wanted so much to have a real princess.

One evening there was a dreadful storm: it
thundered and lightened and the rain poured down
in torrents—it was really quite frightening! Then
there came a knock at the town gate, and the old
king went out and opened it.

It was a princess who stood outside. But my, what a sight she was with the rain and the storm! Her hair and her clothes were running with water: water was running in through the toes of her shoes and out at the heels. But she said she was a real princess.

'Well, we shall soon find out!' thought the old queen to herself. She went into the bedroom, took all the bedclothes off and put a pea on the bottom of the bed. Then she took twenty mattresses and laid them on top of the pea, and then twenty eiderdowns on top of the mattresses.

And there the princess had to spend the night.

In the morning they asked her how she had slept.

'Oh, terribly badly!' said the princess. 'I have hardly shut my eyes the whole night. Heaven knows what there was in the bed. I have been lying on something hard—I am black and blue all over. It's really dreadful!'

And so they could see she was a real princess, because she had felt the pea through twenty mattresses and twenty eiderdowns. No one but a real princess could possibly be so sensitive.

Then the prince married her, for he was now sure that he had found a real princess, and the pea was placed in the art museum, where it can still be seen if no one has taken it.

And that's a true story!

Thumbelina

ONCE upon a time, there was a woman who longed for a little child of her own, but she had no idea where she could get one. So off she went to see an old witch, and she said to her, 'I'd so much like to have a little child. Won't you please tell me where I can get one from?'

'Why, yes, that's no trouble at all,' said the witch. 'You must take this grain of barley—it's not the kind that grows in the farmer's field or that hens eat, either—put it in a flower-pot, and then wait and see what happens.'

'Oh, thank you kindly,' said the woman. She gave the witch a shilling, and then went home and planted the grain of barley. A beautiful big flower sprang up at once—it looked very much like a tulip, but the petals were tightly closed as though it were still in bud.

'What a pretty flower it is!' said the woman, kissing its beautiful red and yellow petals, but just as she kissed it the flower suddenly burst open. You could see it was a real tulip now, but right in the middle of the flower, on a green stool, sat a tiny little girl, delicate and lovely: she was not above an inch tall, and so she was called Thumbelina.

She was given a splendid lacquered walnut-shell as her cradle, blue violet-petals formed her mattress, and a rose-petal her eiderdown. There she slept at night, but during the day she played on the table where the woman had set a plate with a wreath of flowers arranged round the edge so that the stalks lay in the water. On the surface, a large tulip-petal floated, and Thumbelina could sit in it and sail from one side of the plate to the other—she had two white horsehairs to row with. It looked very pretty indeed. She could sing, too, so daintily and charmingly that no one had ever heard the like.

One night, as she lay in her beautiful bed, an ugly old toad came hopping in through the window, where there was a broken pane of glass. The toad looked very big and wet and ugly as she hopped straight down on to the table where Thumbelina lay asleep under her red rose-petal.

'She'd make a nice wife for my son,' said the toad, and so she took hold of the walnut-shell where Thumbelina was sleeping, and hopped off with her through the broken pane down into the garden.

Down at the bottom flowed a big broad stream, and close by the bank the ground was marshy and muddy: here the toad lived with her son. Ugh! he was ugly and repulsive, too, and looked exactly like his mother. 'Croak, croak, cro-ak-ak!' was all he could say when he saw the charming little girl in the walnut-shell.

'Don't talk so loud, or she'll wake!' said the old toad. 'She could still run away from us, for she's as light as swansdown. We'll put her out in the stream on one of the broad water-lily leaves—she's so light and small, it'll seem like an island to her. She won't be able to run away from there, and in the meantime we'll spring-clean the best room down under the mud, and you shall settle down and live there together.'

Out in the stream grew a mass of water-lilies with broad green leaves that looked as if they were floating on top of the water. The leaf that lay farthest out was also the biggest, and the old toad swam out to it and there she put down the walnut-shell with Thumbelina inside.

The poor little thing woke quite early in the morning, and when she saw where she was, she began to cry bitterly, for there was water all round the big green leaf, and she had no way at all of getting to land.

The old toad sat down below the mud, decorating her room with rushes and yellow water-flowers to make it neat and pretty for her new daughter-in-law. Then she swam out with her ugly son to the leaf where Thumbelina was standing. They had come to fetch her pretty bed which they wanted to have ready in the bride's bedroom before she came. The old toad bowed deeply in the water to her and said, 'This is my son you see with me: he's going to be your husband, and the two of you shall have a very nice place to live in down below in the mud.'

But all the son could say was, 'Croak, croak! Cro-ak-ak!'

So they took her pretty little bed and swam away with it, but Thumbelina sat quite alone on the green leaf and wept, for she had no wish to go and live with the horrid old toad, nor to have her ugly son for a husband. The little fishes, swimming down below in the water, had seen the toad and heard what she said, and so they popped their heads up to take a look at the little girl. As soon as they set eyes on her, they saw how lovely she was, and they were very upset that she had to go and live in the mud with the ugly toad. No, that should never happen! Down in the water, they swarmed round the green stalk that held the leaf she was standing on and gnawed it through with their teeth. And so the leaf floated away down the stream with Thumbelina, far away where the toad could not reach her.

Thumbelina sailed along past a great many places, and the little birds sitting in the bushes saw her and sang, 'What a charming little maiden!' The leaf floated farther and farther away with her, and so at last Thumbelina journeyed into a strange country.

A pretty little white butterfly fluttered round her for some time and at last settled on the leaf. It had taken quite a liking to Thumbelina, and she was very happy now, for the toad could no longer reach her and everything was so lovely as she sailed along. The sun was shining upon the water and it looked like gleaming gold. Then she took her sash and tied one end of it round the butterfly and made the other end fast in the leaf so that it now sped along much more swiftly and, of course, took her with it.

Just then a great beetle, a cockchafer, came flying by; he caught sight of her, and in a flash fastened his claws about her slender waist and flew up into a tree with her. But the green leaf floated away down the stream, and the butterfly flew with it, for it was tied to the leaf and could not free itself.

Heavens, how terrified poor Thumbelina was when the cockchafer flew up into the tree with her! But she was most sorry for the beautiful white butterfly she had tied fast to the leaf. If it could not free itself, it would certainly die of hunger. But the cockchafer cared nothing for that. He sat down with her on the largest green leaf in the tree, gave her honeydew from the flowers to eat, and told her she was very pretty, although she was not at all like a cockchafer. Then all the other cockchafers that lived in the tree came to visit her. They looked at Thumbelina, and the young lady cockchafers shrugged their feelers and said, 'But she's only two legs— what a pitiful-looking thing she is!' 'She hasn't any feelers!'

'Her waist is so thin—pooh, she looks just like a human! How ugly she is!' said all the lady cockchafers—and yet Thumbelina was really very pretty. And that's what the cockchafer who had carried her off thought, too, but when all the others said she was ugly, he believed so as well in the end, and would have nothing more to do with her—she could go wherever she would. They flew down from the tree with her and set her upon a daisy. There she wept because she was so ugly that the cockchafers would have nothing to do with her, and yet she was the loveliest little thing you could imagine, as bright and dainty as the most beautiful of rose-petals.

The whole summer through, poor Thumbelina lived quite alone in the great forest. She plaited herself a bed of grasses and hung it under a large dock-leaf to keep the rain off her; she gathered the pollen from the flowers to eat, and she drank the dew that formed every morning on the leaves. And so summer and autumn passed, and winter came—a long, cold winter. All the birds that had sung so beautifully for her flew away, the trees and the flowers withered, and the large dock-leaf she had lived under curled up and nothing was left of it but a shrivelled yellow stalk. She was dreadfully cold, for her clothes were in tatters, and she herself was so delicate and small, poor Thumbelina, that she might have frozen to death. It began to snow, and every snow-flake that fell upon her felt as a whole shovelful might do if it were thrown at one of us, for we are big and she was only an inch tall. So she wrapped herself up in a withered leaf, but it did not warm her, and she shivered with cold.

Just beyond the edge of the forest, which she had now come to, lay a great corn-field, but the corn had long since been carted, and now only the dry bare stubble stood up out of the frozen ground. But it was like a forest for her to go through, and oh, she shivered so much with the cold! And so she came to the field-mouse's door. It was a little hole under the stubble. There the field-mouse lived snug and comfortable with a whole store-room full of corn and a fine kitchen and dining-room. Poor Thumbelina stopped just inside the doorway like a wretched beggar-girl, and asked for a little bit of a grain of barley, for she had had nothing at all to eat for the last two days.

'You poor little thing!' said the field-mouse, for she was a good-natured old field-mouse at heart. 'Come into my warm room and have something to eat with me.'

She liked the look of Thumbelina, and so she said, 'You're welcome to stay with me for the winter, but you must keep my room nice and clean, and tell me stories, for I'm very fond of stories.' And Thumbelina did what the good old field-mouse asked of her, and lived very comfortably.

'We shall soon have a visitor,' said the field-mouse. 'My neighbour comes to visit me every weekday. His house is even more comfortable than mine. He has fine large rooms, and he wears such a lovely black velvet fur—if you could only get him for a husband, you'd be well provided for. But he can't see. You must tell him all the most beautiful stories you know.'

But Thumbelina did not like that idea at all: she had no wish to marry the neighbour, for he was a mole. He came visiting in his black velvet fur—he was very rich and very learned, said the field-mouse. His house was twenty times as spacious as the field-mouse's, and learning he had, too; but he did not care for the sun and the beautiful flowers, and he spoke contemptuously of them because he had never seen them. Thumbelina had to sing for him, and she sang both 'Ladybird, ladybird, fly away home!' and 'The frog he would a-wooing go.' The mole fell in love with her on account of her beautiful voice, but he said nothing, for he was always a very cautious man.

He had recently dug himself a long passage through the earth from his own house to theirs, and he gave the field-mouse and Thumbelina leave to walk there whenever they wished. But he told them not to be afraid of the dead bird that lay in the passage. The bird was quite whole, with its feathers and beak intact: it must have died quite recently at the beginning of the winter, and it now lay buried just where he had made his passage.

The mole took in his mouth a bit of rotten wood, which shines just like fire in the dark, and then went in front to light the long dark passage for them.

When they came to the place where the dead bird lay, the mole set his broad nose against the roof of the passage and pushed the earth up, thus making a big hole which the light could shine through. In the middle of the floor lay a dead swallow with its beautiful wings pressed close against its sides, and its legs and head tucked in under its feathers: the poor bird had obviously died of cold. Thumbelina was so sorry for it. She loved all the little birds that had sung and twittered so beautifully for her the whole summer through. But the mole kicked it with his short legs and said, 'That's one that won't whistle any more. It must be wretched to be born a little bird. Thank Heaven none of my children will ever be one! They've nothing to say for themselves but "tweet-tweet", and in the winter they die of hunger.'

'Yes, as a sensible man, you may well say so,' said the field-mouse. 'What does a bird get for all its "tweet-tweet" when winter comes? It dies of hunger and cold—and yet it's thought so much of.'

Thumbelina said nothing, but when the others had turned their backs upon the bird, she bent down, gently moved aside the feathers that lay over its head, and kissed its closed eyes. 'Perhaps this is the one that sang so beautifully for me during the summer,' she thought. 'What a lot of happiness it brought me, dear, beautiful bird!'

Then the mole stopped up the hole he had made to let the daylight shine through, and went home with the ladies. But that night Thumbelina could not sleep at all. So she got out of bed and plaited a fine large rug of hay. She carried it down and spread it over the dead bird. She tucked some soft cotton-wool she had found in the field-mouse's living-room round the bird's body so that it should lie warm in the cold earth.

'Good-bye, beautiful little bird!' she said. 'Good-bye, and thank you for the lovely song you sang in the summer-time, when all the trees were green and the sun shone warmly on us!' Then she laid her head upon the bird's breast, but as she did so, she was quite startled, for it felt just as if something were knocking inside. It was the bird's heart. The bird was not dead: it lay numb with cold, and now that warmth was returning to it, it showed signs of life again.

In the autumn the swallows all fly away to warmer lands, but if one of them should be delayed, it is so overcome with cold that it falls quite lifeless to the ground. There it remains, and the cold snow covers it over.

Thumbelina had had such a fright that she was actually trembling, for she was only an inch tall and compared with her the bird was a great big creature. But she plucked up her courage, tucked the cotton-wool more closely round the poor swallow, and fetched a mint-leaf which she herself used as a bedspread and laid it over the bird's head.

The next night she crept down to the bird again. It was unmistakably alive, but so weak that it could barely open its eyes for a moment to look at Thumbelina who was standing there with a bit of rotten wood in her hand—for other light had she none.

'Thank you, thank you, pretty little child!' the sick swallow said to her. 'I'm so nice and warm now, I shall soon get my strength back and be able to fly out again into the warm sunshine.'

'Oh,' she said, 'it's so cold outside! It's snowing and freezing. Stay in your warm bed, and I'll look after you.'

Then she brought the swallow some water in a petal, and it drank and told her how it had torn one of its wings on a thorn-bush, and so was unable to fly as fast as the other swallows, who had flown far, far away to warmer lands. At length it had fallen to the ground, but more than that it

33

could not remember and had no idea how it came to be where it was.

The swallow stayed down there all winter, and Thumbelina was kind to it and grew very fond of it, but neither the mole nor the field-mouse was told the least thing about it, for they did not like the poor forlorn swallow.

As soon as spring came and the sun's warmth began to penetrate the earth, the swallow said good-bye to Thumbelina and opened the hole which the mole had made just above its head. It was lovely to feel the sun shining in upon them, and the swallow asked her if she would not go with it: she could sit on its back and they would fly far away into the green forest. But Thumbelina knew it would make the old field-mouse very sad if she left her like that.

'No, I can't do that,' said Thumbelina.

'Good-bye, good-bye, you kind pretty girl!' said the swallow, as it flew out into the sunshine. Thumbelina watched it go, and tears came into her eyes, for she had grown very fond of the poor swallow.

'Tweet, tweet!' sang the bird, and flew off into the green forest.

Thumbelina was very sad. She was never allowed to go out into the warm sunshine; and the corn which had been sown in the field above the field-mouse's house grew up so high that it was like a great dense forest for the poor little girl who was only an inch tall.

'Now during the summer you must get your trousseau ready,' the field-mouse said to her, for their neighbour, the tiresome mole in the black velvet fur, had now asked her to marry him. 'You must have both woollens and linens. You must have something to sit on and lie on when you're the mole's wife.'

Thumbelina had to work hard with her spindle, and the field-mouse hired four spiders to spin and weave for her night and day. Every evening the mole would pay them a visit, and his talk was always the same: when summer came to an end and the sun was no longer shining so dreadfully hot—it was baking the earth as hard as a rock—yes, when summer was over, he would marry Thumbelina. But she was very unhappy about it, for she did not like the tiresome old mole at all. Every morning when the sun rose, and every evening when it set, she would creep out-of-doors, and whenever the wind parted the corn so that she could see the blue sky, she would think how pleasant and beautiful it was outside, and she would wish so hard that she could see her dear swallow again. But it would never come back, for it had flown far away into the beautiful green forest.

When autumn came, Thumbelina had the whole of her trousseau finished.

'In four weeks' time you'll be having your wedding!' the field-mouse said to her. But Thumbelina wept and said she would not have the tiresome old mole.

'Fiddle-de-dee!' said the field-mouse. 'Don't be so obstinate, or I shall bite you with my white teeth. You're going to get a very nice husband indeed. Why, the queen herself has nothing to match his black velvet fur! And he has both a kitchen and a cellar. You should thank God for him.'

And so the wedding-day arrived. The mole had already come to take Thumbelina away; she would have to live with him deep down under the earth, and she would never be able to go out into the warm sunshine because he disliked it. The poor child was very unhappy because she had now to say good-bye to the beautiful sun, which she had still been allowed to look at from the doorway while she was living with the field-mouse.

'Good-bye, bright sun!' she said, stretching her arms up towards it, and she walked a little way from where the field-mouse lived, for the corn was now harvested and there was only the dry stubble left. 'Good-bye, good-bye!' she said, as she threw her small arms round a little red flower that was growing there. 'Give my love to the swallow if you should ever see him again!'

At that very same moment she heard, 'Tweet, tweet!' just above her: she looked up, and there was the swallow flying overhead. It caught sight of Thumbelina and was delighted to see her. She told it how she was being forced to marry the nasty old mole, and how she would then have to live right under the earth where the sun never shone. She could not help weeping at the thought of it.

'The cold winter is coming on now,' said the swallow, 'and I'm flying far away to a warmer land. Will you come with me? You can sit on my back and bind yourself fast with your sash. Then we shall fly away from the nasty mole and his dark home, far away over the mountains to warm lands where the sun shines more beautifully than it does here, where there is always summer and lovely flowers. Do fly away with me, dear little Thumbelina—you saved my life when I lay frozen in the dark cellar under the earth.'

'Yes, I'll go with you,' said Thumbelina, and she seated herself upon the bird's back, with her feet resting on its outstretched wings, and tied her belt firmly to one of its strongest feathers. And so the swallow flew high up into the air, over forest and lake, high up over the great mountains where snow always lies. Thumbelina froze in the cold air, but she crept under the bird's warm feathers, and only popped her little head out to see all the beautiful scenery below her.

So they came to the warm lands where the sun shines much more brightly than it does here. The sky seemed twice as high, and along the ditches and hedges grew the loveliest green and blue grapes. Oranges and lemons hung in woods scented with myrtle and mint, and along the country lanes lovely children ran and played with great brightly coloured butterflies. But the swallow flew farther still, and everything grew more and more beautiful. Under magnificent green trees, by a blue lake, stood a palace built in ancient times of shining white marble, with vines entwined round its tall pillars. Right on top of the pillars were many swallows' nests, and in one of them lived the swallow that carried Thumbelina.

'Here's my house,' said the swallow. 'But if you will choose yourself one of those lovely flowers growing down there, I will set you down on it, and you will live as happily there as you could wish.'

'That would be lovely!' she said, clapping her little hands.

A great white marble column lay fallen on the ground. It was broken into three pieces, and between them grew the most beautiful white flowers. The swallow flew down with Thumbelina and set her upon one of the broad petals: but what a surprise she had!—there, in the middle of the flower, sat a little man, clear and transparent like glass. He had the prettiest golden crown upon his head and the loveliest bright wings on his shoulders; and he himself was no bigger than Thumbelina. He was the guardian-spirit of the flower. In every flower there lived a little man or woman just like him, but he was king over them all.

'Oh, how beautiful he is!' whispered Thumbelina to the swallow. The little prince was quite frightened of the swallow, for it was such an enormous bird by the side of this dainty little creature, but when he saw Thumbelina he was overjoyed, for she was quite the most beautiful girl he had ever seen. And so he took the golden crown off his own head and placed it upon hers. He asked her what she was called, and he said that if she would be his wife, she should be queen over all the flowers. Well, that was a husband indeed!—quite a different matter from the toad's son and the mole with the black velvet fur! And so she said yes to the handsome prince. From every flower appeared a little lady or gentleman, so beautiful and dainty it was a joy to look at them; and every one brought Thumbelina a present, but the best of all was a pair of beautiful wings from a large white fly. They were fastened upon Thumbelina's shoulders, and then she, too, could fly from flower to flower. They were all so full of joy, and the swallow sat up above in its nest and sang for them as well as it could, but it was very sad at heart, for it was so fond of Thumbelina that it wanted never to be parted from her.

'You cannot be called Thumbelina,' the flower-sprite said to her. 'It's an ugly name, and you are so beautiful. We shall call you Maia.'

'Good-bye, good-bye!' said the swallow, and flew away from the warm lands once more, far away back to Denmark. There it had a little nest by the window of the room where the man who can tell tales lives, and it sang its 'tweet, tweet!' for him, and that's how we come to know the whole story.

The Travelling Companion

POOR Hans was very sad, for his father was seriously ill and was not expected to live. The two of them were quite alone in their little room. The lamp on the table was on the point of burning out, and it was quite late in the evening.

'You've been a good son, Hans', his sick father said. 'I'm sure Our Lord will help you to get on in the world.' He looked at him with a serious and gentle expression in his eyes, drew a deep breath, and died, looking just as if he were asleep. But Hans wept, for now he had no one at all in the whole world, neither father nor mother, brother nor sister. Poor Hans! He leaned kneeling against the side of the bed, weeping bitterly; but at last his eyes closed and he fell asleep with his head on the hard wooden bed-frame.

Then he dreamt a strange dream: he saw the sun and the moon bow before him, and he saw his father lively and well again, and heard him laugh as he always used to laugh when he was really pleased. A girl with a golden crown on her beautiful long hair held out her hand to Hans, and his father said, 'Look what a lovely bride you have, the loveliest in the whole world!' Then he woke—the dream had gone, his father lay dead and cold in the bed, and there was no one at all with them—poor Hans!

A week later the dead man was buried. Hans walked close behind the coffin, and as the earth was thrown down upon it, he felt as if his heart would break. But the hymn they sang round about him, and the sun shining brightly on the green trees comforted him.

He cut a large wooden cross to put on his father's grave, and when he brought it along in the evening he found the grave made neat with sand and decorated with flowers: their neighbours had done this, for they thought very highly of his dear father who was now dead.

Early the next morning Hans packed his little bundle and tucked away in his belt the whole of his inheritance—some fifty silver crowns and a couple of shillings—and with that he set out to make his way in the world. But first he walked over to the churchyard to his father's grave, said the Lord's Prayer, and added, 'Good-bye, dear father! I will always try to be good, so you needn't be afraid to pray to God that all may go well with me.'

As Hans walked along through the fields, all the flowers looked fresh and lovely in the warm sunshine, and they nodded in the wind as if they would say, 'Welcome to the green fields! Isn't it delightful out here?' But Hans turned round once again to look at the old church where he had been baptized as a little child and where he had gone to church every Sunday with his father and joined in the hymns.

Then he saw the church goblin with his little red pointed hood standing high up in one of the holes in the tower and shading his face with his bent arm to keep the sun out of his eyes. Hans nodded good-bye to him, and the little goblin waved his red hood, put his hand upon his heart, and blew him kiss after kiss to wish him good luck and a happy journey.

Hans thought about all the beautiful things he was going to see in the fine wide world, and he went on and on, farther than he had ever been before. He knew neither the towns he passed through nor the people he met—he was now among strangers, far away from home.

The first night, he lay down to sleep in a haystack in the fields, for other bed had he none. But he thought it was just lovely—the king himself couldn't have a finer bed! The whole field, with the stream and the haystack and the blue sky overhead, was just like a beautiful bedroom. The green grass dotted with little red and white flowers was his carpet, the elderberry bushes and the hedges of wild rose his pots of flowers, and for his wash-basin he had the stream with its clear fresh water, where the rushes bowed to him and said good evening and good morning. The moon was a fine great night-light hanging high up under the blue ceiling—and it couldn't set fire to the curtains! Hans could sleep quite peacefully there, and he did, too. He did not wake up until the sun rose and all the little birds were singing, 'Good morning, good morning! Aren't you up yet?'

It was Sunday and the bells were ringing for church. People were passing by on their way to early morning service. Hans went with them, sang a hymn and listened to the sermon. He felt just as if he were back in his own church where he had been christened and had sung hymns with his father.

Outside in the churchyard were a great many graves, some of them overgrown with tall grass. Then Hans thought of his father's grave. Now that he was no longer there to weed it and keep it looking nice, it, too, might come to look like these. So he knelt down and pulled up the grass, righted the wooden crosses which had fallen down, and put back again in their proper place the wreaths which the wind had torn away from the graves; and as he worked, he hoped perhaps someone would do the same thing for his father's grave now that he could no longer do it himself.

Outside the churchyard gate stood an old beggar leaning on his crutch. Hans gave him the two odd shillings he had and then went on his way, full of happiness and joy, out into the wide world.

Towards evening a dreadful storm sprang up. Hans hurried along to find a roof for his head, but he was very soon overtaken by the dark. At last, however, he reached a little church which stood all alone on top of a hill. By good luck the door was on the latch, and so he slipped in: he would stay there until the storm had passed.

'I'll find a corner where I can sit down,' he said. 'I'm quite tired and I can do with a little rest.' So he sat down, clasped his hands together and said his prayers; and before he knew what had happened he had fallen asleep and was dreaming, while it thundered and lightened outside.

When he woke again it was the middle of the night, but the weather had cleared and the moon was shining in through the windows. In the middle of the church there stood an open coffin with a dead man inside still awaiting burial. Hans was not at all frightened, for he had a good conscience and he knew that the dead harm nobody—only the living cause harm. Two evil men were standing close by the coffin which had been placed there in the church before being laid in the grave. They were intent upon doing the dead man harm. They would not let him lie peacefully in his coffin and were about to throw his poor body outside the church door.

'Why are you doing that?' asked Hans. 'That's wicked

and sinful! Let him sleep peacefully in the Name of Jesus!'

'Rubbish!' said the two evil men. 'He's swindled us. He owes us money which he couldn't pay back, and now, on top of that, he's died and we shan't get a penny. So we intend to get our own back on him—he shall lie like a dog outside the church door!'

'I've only fifty silver crowns,' said Hans. 'It's the whole of my inheritance, but I'll give it to you willingly if you'll promise me faithfully to leave the poor dead man in peace. I shall get on all right without the money; I've strong healthy limbs and Our Lord will always help me.'

'All right,' said these two loathsome fellows, 'since you're willing to pay his debts, we won't do him any harm, you can be sure of that!' So they took the money Hans gave them, laughed openly at his good nature, and went their way. But Hans set the corpse to rights again in the coffin, folded its hands over its breast, said good-bye and went contentedly upon his way through the great forest.

Round about him, wherever the moon shone through between the trees, he saw the little elves at play. They were not in the least disturbed by his presence for they knew he was good and innocent, and it is only wicked folk who are not allowed to catch sight of them. Some of them were no bigger than a finger, and their long golden hair was pinned up with combs of gold. Two by two they see-sawed up and down on the great dew-drops that lay upon the leaves and the tall grasses. From time to time a dew-drop would roll off, and then they would fall down among the long grass-stalks amid the laughter and noise of the other little elves—they thought that was great fun! They sang, and Hans was able to pick out quite clearly the verses he had learnt as a little boy. Large many-coloured spiders with silver crowns upon their heads were spinning long hanging bridges and palaces from one hedge to the other, and when the fine dew fell upon them, they looked like glass in the clear moonshine. And so it went on until sunrise, when the elves crept into flower-buds and the wind caught their bridges and castles which drifted away in the air like large cobwebs.

Hans had just left the forest when a man's strong voice shouted behind him, 'Hallo, my friend! Which way does your journey take you?'

'Out into the wide world,' said Hans. 'I've neither father nor mother. I'm a poor lad, but Our Lord will help me!'

'I'm going out into the wide world, too,' said the stranger. 'Couldn't we be company for one another?'

'All right,' said Hans, and off they went together. They soon grew very fond of one another, for they were both good-hearted people. But Hans noticed that the stranger was much wiser than he was—he had been nearly all over the world, and could talk about absolutely everything.

The sun was already high when they sat down under a great tree to eat their lunch. Just at that moment an old woman came along. She was very old and quite bent and supported herself with a stick. On her back she carried a bundle of kindling which she had gathered in the forest. Her apron was tucked up and Hans saw three large bundles of bracken and willow-twigs sticking out of it. She was nearly up to them when her foot slipped and she fell with a loud shriek. The poor old woman had broken her leg.

Hans wanted to carry her home at once, but the stranger opened his bag, took out a jar, and said he had an ointment there which would make her leg quite well again right away, so that she would be able to walk home by herself, just as if she had never broken her leg at all.

But in return he wanted her to make him a present of the three bundles of sticks she had in her apron.

'It's a steep price to pay!' said the old woman, nodding her head rather oddly. She did not much like the idea of parting with her sticks, but it wasn't any fun lying there with a broken leg either. So she gave him the sticks, and as soon as he had rubbed the ointment on her leg, the old woman got up and off she walked better than ever. The ointment was as powerful as that—but it's not to be had from the chemist's!

'What do you want those sticks for?' Hans asked his travelling companion.

'They're three fine birch-brooms,' he said. 'I've taken a fancy to them—I'm a queer sort of fellow!'

So they went on a good bit farther.

'My, look at that storm gathering!' said Hans, pointing straight ahead. 'Look at those dreadful thick clouds there!'

'No, they're not clouds,' said his travelling companion. 'They're mountains, fine high mountains, where you climb right up above the clouds into clear fresh air. It's magnificent, believe me! We shall reach them tomorrow and be well on our way into the wide world.'

The mountains were not, however, as near as they looked. It took a whole day before they reached them, and there the dark forests grew right up towards the sky and there were great masses of rock as big as whole towns. It was going to be a difficult enough journey to cross right over them, and so Hans and his travelling companion went into an inn where they could have a good rest and gather strength for the next day's march.

Down in the great tap-room of the inn a large crowd of people had gathered, for there was a man there with a puppet-show. He had just set his little theatre up and people were sitting round to see his play. Right in front a stout old butcher had taken his seat, the best in the room, and his great bulldog—an ugly-looking brute it was—sat by his side staring open-eyed like the rest of them.

The play now began. It was a fine play with a king and a queen who sat on plush thrones and had golden crowns upon their heads and wore robes with long trains, which, of course, they could well afford. Very charming wooden puppets with glass eyes and great moustaches stood by all the doors, and opened and shut them to let fresh air into the room. It was a delightful play and not a bit sad, but just as the queen rose to walk across the floor the great bulldog, at a moment when the stout butcher was not holding him, bounded right into the theatre, and—God knows what he was thinking of!—seized the queen by her slender waist so that it went 'Crick-crack!' It was quite terrifying!

The poor puppet-man, who, of course, was performing the whole play, was frightened out of his wits, and very upset over his queen, for she was the most charming puppet he had, and now that dreadful bulldog had bitten her head off. But afterwards when the people had gone, the stranger who had come with Hans said he would very soon put her to rights again. He took out his jar and smeared the puppet with the ointment that had cured the poor old woman when she had broken her leg. As soon as the puppet was smeared with it, it was whole again at once—yes, and more than that, it could now move all its limbs by itself. There was no need at all to pull its strings—the puppet was just like a living person, with the one exception that it could not talk. The man who owned the little puppet-theatre was delighted: he would no longer have to hold the puppet now that it could dance so well by itself. And none of the others could do that!

When at last night came, and everybody in the inn had gone to bed, they heard someone sighing deeply. It was such a dreadful noise and it went on so long that they all got up to see who it could be. The man who had given the play went down to his little theatre, for that was where the sighs were coming from. All the wooden puppets lay tumbled together, the king and all his guards, and it was they who were sighing so pitifully, staring with their great glass eyes, because they did so want to be smeared with a little of the ointment like the queen, so that they, too, would be able to move by themselves. The queen threw herself down on her knees and held out her lovely golden crown. 'Take it,' she begged, 'but do smear my husband and my court!' Then the poor man that owned the show and all the puppets could not help bursting into tears, for he was really sorry for them. He immediately promised Hans' travelling companion that he would give him all the money he took at the next evening's performance if he would only smear four or five of his finest puppets. But the travelling companion said he desired nothing but the great sabre that the puppet-man wore at his side. When he was given it he annointed six puppets who immediately began to dance, and so charmingly that all the girls—the real live girls—who saw them joined in too. The coachman and the kitchen-maid were dancing, the waiter and the chamber-maid, all the people who were staying there, and even the coal-shovel and the fire-tongs, but these two fell over just as they did their first hop—it was a merry night indeed!

The next morning Hans and his travelling companion left them all, and made their way up the high mountains and through the great forest of firs. They climbed so high that at last the church towers far down beneath them looked like small red berries dotted on the green landscape below, and they could see for miles and miles, far, far away where they had never been. Never before had Hans seen so much of the beauty of this world at one time. The sun shone warmly from the clear blue sky, and he heard the huntsmen too, blowing their horns so beautifully in among the mountains. Tears of joy came into his eyes, and he could not help saying, 'Dear, kind Lord, I could kiss you for being so good to us all, and for giving us all the loveliness there is in the world!'

His companion, too, stood with folded hands and looked out over the woods and villages in the warm sunshine. At that moment they heard a wonderful sound ring out above their heads. They looked up: a great white swan was hovering in the air. It was very beautiful and it sang as they had never before heard any bird sing. But the swan grew weaker and weaker, bowed its head and sank slowly down to their feet, where it lay dead in all its beauty.

'Two such lovely wings,' said Hans' companion, 'big and white like these, are worth some money. I'll take them with me. Now you see what a good thing it was I got this sabre!'

And so with one stroke he cut both the wings off the dead swan and kept them.

They now travelled many, many miles onwards over the mountains, until at last they saw in front of them a great city with more than a hundred towers shining like silver in the sunshine. In the middle of the city was a magnificent marble palace roofed with red gold, and there lived the king.

Hans and his companion did not want to go into the city right away, and so they put up at an inn just outside where they could tidy themselves up, for they wanted to look smart when they appeared in the streets. The innkeeper told them that the king was a good sort of man who never did anybody any harm, but his daughter—God help us, there was a wicked princess, if ever there was one! She was beautiful enough—no one was ever as pretty and attractive as she was—but what was the good of that? She was a cruel, wicked witch, and it was all because of her that so many handsome princes had lost their lives. She had given everybody leave to court her. Anyone could come, prince or beggar for all she cared; and all he had to do was guess three things she asked him. If he could do that, she would marry him and he would be king over the whole land when her father died; but if he could not guess the right answers, she had him hanged or beheaded, so cruel was she.

Her father, the old king, was deeply distressed by it all, but he could not forbid her evil ways, for he had once said he would not interfere with her suitors, she was to do as she liked about them. Every prince who came and made his guesses to win the princess failed, and so he was either hanged or had his head cut off, though he had been warned in good time and could have given up the idea of wooing her. The old king was so upset by all the sorrow and misery that was caused that he spent a whole day every year upon his knees with all his soldiers praying that the princess might grow good, but she showed no sign of doing so. The old women who drank brandy coloured it black before they drank it; it was their way of showing their sorrow, and more than that they could not do.

'What a dreadful princess!' said Hans. 'She ought to have a good thrashing; it would do her good. If only I were the old king she'd be well and truly whipped!'

At that moment they heard the people outside shouting 'Hurray!' The princess was passing by, and she really was so lovely that everyone forgot how bad she was, and so they shouted 'Hurray!' Twelve lovely maidens, all in white silk dresses, with golden tulips in their hands, rode by her side on coal-black horses. The princess herself had a snow-white horse decked with diamonds and rubies, her riding-habit was of pure gold, and the whip she carried in her hand looked like a ray of sunlight. The golden crown upon her head glittered like small stars from heaven, and her cloak was embroidered all over with thousands of butterfly wings – and yet she was far more beautiful than all her clothes put together.

When Hans caught sight of her he went as red in the face as a drop of blood, and he could hardly utter a word; for the princess looked exactly like the lovely girl with the golden crown he had dreamt of the night his father died. He thought she was very beautiful, and he could not help loving her deeply. It couldn't possibly be true, he thought, that she was a wicked witch who had people hanged or beheaded when they couldn't guess what she asked them. 'Everyone is free to woo her, even the poorest beggar—I must go up to the palace! I can't help it!'

They all said he shouldn't do it; he would certainly fare no better than the rest. His travelling companion, too, advised him not to go, but Hans, certain that all would be well, brushed his shoes and jacket, washed his hands and face, combed his beautiful fair hair, and then walked all alone into the city and up to the palace.

'Come in!' said the old king, when Hans knocked at the door. Hans opened the door, and the old king, in dressing-gown and embroidered slippers, came to meet him. He had his golden crown upon his head, his sceptre in one hand and his golden orb in the other. 'Wait a minute!' he said, tucking the orb under his arm in order to shake Hans by the hand. But as soon as he heard he was another suitor come to woo the princess, he burst into such a flood of tears that both sceptre and orb fell to the floor, and the poor old king had to dry his eyes on his dressing-gown.

'Don't do it!' he said. 'You'll come to a bad end like all the rest. Just look here!' He took Hans out into the princess's pleasure-garden. It was a ghastly sight. From every tree hung three or four princes who had wooed the princess but had not been able to guess the things she had asked them. Every time the wind blew, all the bones rattled so that the little birds were frightened and never dared go into the garden; the flowers were all tied up to human bones, and skulls stood grinning in the flower-pots. It was a fine garden for a princess!

'Now you can see!' said the old King. 'You'll end just like all the others you see here. You'd do far better to drop the idea. You make me really unhappy—I take it all very much to heart!'

Hans kissed the good old king's hand and said all would be well, because he loved the princess so much.

Just then the princess herself came riding into the palace courtyard with all her ladies, and so they went over to her and said good-day. She was lovely, there was no doubt of it, and as she gave Hans her hand he fell even more deeply in love with her than before—she couldn't possibly be the cruel witch that everybody said she was. They went up into the hall, and the little pages offered them sugared fruits and humbugs, but the old king was so upset he couldn't eat anything at all, and besides, the humbugs were too hard for him.

It was now decided that Hans should come up to the palace again the next morning, when the judges and the whole Court would assemble to hear whether he would succeed in guessing what the princess was thinking of. If he guessed correctly he would have to appear twice again before them; but so far no one had ever guessed right the first time, and so they had all had to lose their lives.

Hans was not at all worried about what would happen to him. He knew only that he was happy, and he thought of nothing but the lovely princess. He was quite certain that God would help him, though how, he neither knew nor thought about. He danced all the way down the road on his way back to the inn where his travelling companion was waiting for him.

Hans could not stop telling him how charming the princess had been to him and how lovely she was. He was already longing for the next day, when he would go to the palace again and try his luck at guessing.

But his companion shook his head and was overcome with grief. 'I'm very fond of you,' he said. 'We could have spent a long time together yet, and now I must lose you already! My poor, dear Hans, I could cry, but I won't spoil your happiness on what is, perhaps, the last evening we shall have together. We'll be merry, really merry. Tomorrow, when you are gone, I shall be free to cry.'

The news had already spread through the town that a new suitor had come to woo the princess, and there was great grief everywhere. The theatre was closed, all the women who sold cakes tied black crape round their sugar-pigs, and the king and the clergy were on their knees in church—there was such great grief because no one believed that Hans could do any better than all the other suitors had done.

Late in the evening Hans' travelling companion made a great bowl of punch, and said they should now be really merry and drink the princess's health. But when Hans had drunk two glasses he became so sleepy that he could not keep his eyes open, and he could not stop himself falling asleep. His companion lifted him gently out of his chair and put him to bed. Then, when it was quite dark, he took the two great wings he had cut off the swan and bound them firmly to his shoulders; he put in his pocket the biggest of the bundles of twigs he had got from the old woman who had fallen and broken her leg; and then he opened the window and flew out over the city straight to the palace, where he settled himself in a corner high up under the window of the princess's bedroom.

The whole city was quite still, and then a quarter-to-twelve struck—the window opened, and the princess, in a large white cloak flew with long black wings away over the city to a great mountain. But the travelling companion made himself invisible so that she should not see him at all, and, flying behind her, thrashed the princess with his twigs so that he actually drew blood where he struck her. How swiftly they sped through the air! The wind took her cloak, spreading it out on all sides like a great sail with the moon shining through it.

'Oh, how it's hailing, how it's hailing!' said the princess with every stroke of the twigs, but she got no more than she deserved. Then at last she came to the mountain and knocked on it. There was a rolling like thunder as the mountain opened and the princess passed inside. Hans' travelling companion slipped in beside her, for no one could see him since he was invisible. They went down a great long passage where the walls sparkled, and thousands of glowing spiders ran up and down them shining like fire. Then they entered a great hall built of silver and gold, with flowers as big as sunflowers shining red and blue from the walls: but no one could pick them, for the stalks were vile poisonous snakes and the flowers were the fire that flamed from their mouths. The whole ceiling was covered with shining glow-worms and sky-blue bats beating their thin wings. It was an amazing spectacle. In the middle of the floor there was a throne borne on the skeletons of four horses with harnesses of red fire-spiders; the throne itself was of milk-white glass, and the cushions to sit on were small black mice biting each other's tails. Over the top of it was a canopy of rose-red cobweb set with the loveliest little green flies that shone like precious stones. In the middle of the throne sat an old troll with a crown upon his ugly head and a sceptre in his hand. He kissed the princess on her forehead and made her sit beside him on his throne.

Then the music began. Large black grasshoppers played the jew's harp, and the owl beat his belly since he had no drum. It was a strange concert. Tiny little goblins with will-o'-the-wisps on their hoods danced round the hall. No one could see Hans' travelling companion, who had placed himself just behind the throne, and heard and saw everything. The courtiers, who had now come in, were very fine and distinguished-looking, but anyone who looked at them hard could not have failed to notice what they really were. They were nothing but broomsticks with cabbages on top for heads, which the troll had brought to life with a spell and provided with finely embroidered clothes. But it did not matter in the least—they were only there for show.

When they had danced for some time the princess told the troll she had a new suitor, and asked him what she should think of to ask the suitor the next morning when he came up to the palace.

'Now, listen to me,' said the troll, 'and I'll tell you! You must find something very ordinary, then he'll never hit upon it. Think of one of your shoes. He won't guess that. Then have his head cut off, but don't forget, when you come and see me again tomorrow night, to bring me his eyes—I want them to eat!'

The princess curtsied deeply and said she would not forget the eyes. Then the troll opened the mountain and she flew home again, but the travelling companion went with her and gave her such a thrashing with the twigs that she sighed deeply over the fierce hail-storm and hastened for all she was worth to get back through her window into her bedroom again. But the travelling companion flew back to the inn where Hans was still asleep, undid his wings and then lay down on the bed, too, for I've no doubt he was very tired.

It was quite early in the morning when Hans woke up. His companion got up too, and told him that during the night he had dreamt a very strange dream about the princess and one of her shoes. It was, of course, what he had heard the troll say inside the mountain, but he did not want to tell Hans anything about that, and so he only begged him to answer that she had been thinking of one of her shoes.

'I might just as well say that as anything else,' said Hans. 'Maybe what you've dreamt is the right answer, for I've always believed Our Lord would help me. But all the same, I'll say good-bye to you, for if I should guess wrong, I shall never see you again.'

So they kissed one another and Hans went into the city and up to the palace. The hall was quite full of people. The judges were sitting in their arm-chairs and had soft down cushions behind their heads because they had so much to think about. The old king was standing and drying his eyes on a white handkerchief. Then the princess came in, looking even lovelier than the day before. She greeted them all very kindly, but to Hans she gave her hand and said, 'Good morning to you!'

And now the time had come for Hans to guess what she was thinking of. How friendly she seemed as she looked at him! Yet as soon as she heard him say, 'One of your shoes!' she grew deathly white in the face and trembled all over. But there was nothing she could do about it, for he had guessed right.

Hurray! How glad the old king was. He turned head over heels with excitement, and everybody clapped him, and they clapped Hans, too, who had now guessed the answer to the first question.

The travelling companion beamed with joy when he learnt how well it had all gone; but Hans clasped his hands together and thanked God, who, he was sure, would help him again the next two times. He had to make his second guess the very next day.

The evening passed just like that of the day before. When Hans had fallen asleep, his companion flew away after the princess to the mountain, and gave her an even harder thrashing than the previous time, for he had now taken two bundles of twigs with him. No one could see him and he heard everything that passed. The princess was to think of her glove. And he told Hans that as if it had been a dream.

So Hans was able to guess right again, and there was great rejoicing in the palace. The whole Court turned head over heels, just as they had seen the king do the first time. But the princess went and lay down on the sofa, and refused to say a single word. And now everything depended upon whether Hans could guess right the third time. If all went well he would win the lovely princess and inherit the kingdom when the old king died; if he guessed wrong, he would lose his life and the troll would eat his handsome blue eyes.

That evening Hans went to bed early, said his prayers, and then slept quite peacefully. Meanwhile his travelling companion strapped his wings upon his back, buckled his sword by his side and took all three bundles of sticks with him, and then he flew to the palace.

It was a pitch-black night, and there was such a gale blowing that tiles flew off the houses, and the trees in the gardens where the skeletons hung swayed like reeds in the wind. Lightning flashed every moment, and the thunder rolled as if there were but one single peal that lasted the whole night through. Then the window was flung open and the princess flew out; she was deathly pale, but she laughed at the storm as though she thought it was not strong enough for her, and her white cloak whirled round in the air like a great sail. The travelling companion thrashed her so hard with his three bundles of sticks that her blood dripped upon the earth below and in the end she could hardly fly any farther. But she reached the mountain at last.

'You should see how it's hailing and blowing!' she said. 'I've never been out in such weather!'

'You can have too much even of a good thing,' said the troll. Then she told him that Hans had guessed right the second time as well; if he should do the same again in the morning he would have won, and she would never more be able to visit the troll in the mountain or practise witchcraft; and so she was quite downcast.

'He mustn't be able to guess,' said the troll. 'I'm sure I shall think of something he'll never guess – or else he must be a greater wizard than I am. But now let's be merry!' And so he took the princess by both hands, and they danced round with all the little elves and will-o'-the-wisps that were in the room; the red spiders ran up and down the walls as merrily as the rest of them, and the fire-flowers seemed to send out showers of sparks. The owl beat his drum, the crickets whistled, and the black grasshoppers blew upon their jew's harps. It was a lively ball!

Then when they had danced quite a long time, the princess had to go home for fear she might be missed at the palace. The troll said he would go with her so that they could be together a little longer.

Then they flew away into the storm, and the travelling companion wore out his three bundles of sticks upon their backs. Never had the troll been out in such a hail-storm! Outside the palace he said good-bye to the princess, at the same time whispering to her, 'Think of my head!' But the travelling companion heard quite clearly what he said, and at the very moment when the princess slipped through the window into her bedroom and the troll was about to turn back, he caught hold of him by his long black beard, and struck his ugly troll's head right off his shoulders with his sword. He threw the body out to the fish in the lake, but the head he just dipped in the water: then he tied it up in his silk handkerchief and took it home with him to the inn, where he lay down to sleep.

The next morning he gave Hans the handkerchief, but said he must not untie it until the princess asked him what she was thinking of.

There were so many people in the great hall of the palace that they were packed together like radishes tied in a bundle. The judges were sitting in their seats with their soft head-cushions, and the old king was wearing a new suit of clothes. The golden crown and sceptre had been polished and were looking magnificent. But the princess was quite pale, and wore a coal-black dress as if she were going to a funeral.

'What am I thinking of?' she asked Hans, and he immediately untied the handkerchief and was quite frightened himself when he saw that frightful troll's head. Everybody shivered, for it was gruesome to look at, but the princess sat like a statue, unable to say a single word. At last she rose and gave Hans her hand, for he had, of course, guessed right. She looked at no one, but sighed deeply and said, 'You are now my lord! We will be married this evening!'

'I approve of that!' said the old king. 'That's just what we'll do!'

Everyone cried 'Hurray!', the Guards' band played in the streets, the bells rang, and the cake-women took the black crape off their sugar-pigs, for now there was great rejoicing. Three oxen, roasted whole and stuffed with ducks and hens, were put in the middle of the market-place, and anyone could cut himself off a slice; the finest wine gushed from the fountains; and if you bought a penny roll at the baker's, you got six big buns thrown in—and buns with raisins in, too!

In the evening the whole city was illuminated, and the soldiers let off their cannon and the boys their crackers, and there was eating and drinking, clinking of glasses and merry-making up at the palace, and all the noble gentlemen and the lovely young ladies danced together: you could hear them a long way off singing:

'Here are so many pretty maidens,
And they want to swing around
To the rhythm of the drum.
Pretty maiden, just turn round,
Let us dance and stamp our feet
Till the soles fall off our shoes!'

But the princess was still a witch, of course, and cared nothing at all for Hans. The travelling companion did not forget that, and so he gave Hans three feathers from the wings of the swan and a little flask with some drops in it. He told Hans he must have a large tub filled with water put by the bride's bed, and as the princess was climbing into bed he must give her a little push so that she would fall into the water. Then after he had first thrown the feathers and the drops in he must duck her under three times, and then she would be freed from her witchcraft and grow to love him very much.

Hans did everything as his companion had advised him. The princess shrieked quite loudly as he ducked her under the water, and she floundered about under his hands in the form of a large coal-black swan with glinting eyes. When she came up again for the second time, she had turned into a white swan with just a single black ring round her neck. Hans prayed fervently to Our Lord and let the water run over the bird for the third time, and it changed back immediately into the lovely princess. She was even more beautiful than before, and she thanked him with tears in her lovely eyes, for freeing her from the spell.

The next morning the old king came with the whole of the Royal Household, and their congratulations continued far into the day. Last of all came Hans' travelling companion with his staff in his hand and his haversack on his back. Hans kissed him over and over again, and told him he must not go away but stay with him, since he owed him all his happiness. But his companion shook his head and said very kindly and gently, 'No, my time is up now. I have only paid my debt. Do you remember the dead man those wicked men wanted to harm? You gave them all you had so that he could lie quietly in his grave. I am that dead man!'

At that very moment he was gone.

The wedding-feast lasted a whole month. Hans and the princess loved one another very much. The old king lived for many a happy day, and let his grandchildren ride-a-cock-horse on his knee and play with his sceptre. And after his death Hans became king over the whole country.

The Little Mermaid

FAR out to sea the water is as blue as the petals of
the loveliest cornflower, and as clear as the purest
glass, but it is very deep, deeper than any anchor-
chain can reach, and many church towers would
have to be put one on top of another to reach from
the bottom to the surface of the water. Down there
live the mer-folk.

Now you must not think that there is nothing
there but the bare white sandy bottom; oh no, the
most wonderful trees and plants grow there, and
their stems and leaves are so supple that with the
least movement of the water they stir like living
things. All kinds of fish, large and small, flit among
the branches just as birds do in the air up here. In

the very deepest place of all lies the mer-king's palace. Its walls are coral, and its long pointed windows the clearest amber, and its roof is made of mussel-shells that open and shut with the movement of the water—it looks most beautiful, for in each shell lies a shining pearl, and a single one of them would be the chief beauty in a queen's crown.

The mer-king down there had been a widower for many years, and his old mother kept house for him; she was a wise woman, but proud of her royal rank, and so she always went about with twelve oysters on her tail, while other high-born mer-folk might have only six. Otherwise she deserved high praise, especially because she was so fond of the little mer-princesses, her granddaughters. They were six lovely children, and the youngest was the most beautiful of them all— her skin was as clear and pure as a rose-petal, her eyes as blue as the deepest lake, but like all the rest, she had no feet and her body ended in a fish's tail.

All day long they could play down there in the palace, in the great halls where living flowers grew out of the walls. When the great amber windows were opened, the fish would swim in to them, just as the swallows will fly in when we open our windows, but the fish would swim right up to the little princesses, eat out of their hands and let themselves be stroked.

Outside the palace there was a great garden with fiery-red and dark-blue trees, their fruit shining like gold and their flowers like burning fire, with ever-moving stems and leaves. The soil itself was the finest sand, but blue like sulphur-flames. Over everything down there lay a wonderful blue glow; you would think that, instead of being on the bottom of the sea, you were standing high up in the air with nothing but sky above and below you. In clear calm weather you could see the sun looking like a purple flower with all that light streaming out from its centre.

Each of the little princesses had her own little plot in the garden, where she could dig and plant whatever she liked.

One gave her flower-bed the shape of a whale, another thought hers would look better in the form of a little mermaid, but the youngest made hers quite round and would only have flowers that shone red like the sun. She was a strange child, quiet and thoughtful. When the other sisters decorated their gardens with the most wonderful things they had got from wrecked ships, she would only have, apart from the rose-red flowers which looked like the sun high up above her, a beautiful marble statue, a handsome boy, carved out of clear white stone and brought down to the bottom of the sea with the wreck of a ship. She planted by it a rose-red weeping-willow which grew magnificently and hung its fresh branches right over the statue down to the blue sand of the ground, where its shadow showed violet and was in constant motion like its branches—the top of the tree and its roots seemed always to be playing at kissing one another.

She found her greatest delight in hearing about the world of men up above; her old grandmother had to tell her all she knew about ships and cities, men and beasts. What she thought especially wonderful and beautiful was that up on earth the flowers had a sweet scent, for that they did not have on the bottom of the sea, and the woods were green and the fish you could see there among the branches could sing so loudly and beautifully it was a joy to hear them—grandmother called the little birds fish, for otherwise they would not have been able to understand her, for they had never seen a bird.

'When you reach your fifteenth birthday,' said grandmother, 'you shall be allowed to go up out of the sea and sit in the moonshine on the rocks and watch the great ships sailing by, and you shall see woods and cities, too!'

During that year one of the sisters was fifteen, but the others—well, each was a year younger than the next, and so the youngest of them all still had five whole years to wait before she dared swim up from the bottom of the sea and see what it looks like where we live. But each one promised to tell the next what she had seen and found most lovely during her first day on earth, for their grandmother had not told them nearly enough—there was so much they wanted to know about.

None of them was so full of longing as the youngest and she was the very one that had the longest to wait and was so quiet and thoughtful. Many a night she stood by the open window and looked up through the dark-blue water where the fish were waving their fins and tails. She could see the moon and the stars, and though they shone quite palely, they looked much larger through the water than they do to our eyes; and when it looked as if a black cloud were gliding across below them, she knew it was either a whale swimming over her or else a ship with many men on board—men who, of course, had no idea that a lovely little mermaid was standing down below and stretching her white hands up towards their keel.

And now the eldest princess was fifteen years old and could rise up above the surface of the sea.

When she came back, she had hundreds of things to tell them, but the loveliest of all, she said, was to lie in the moonshine on a sandbank in the quiet of the sea, and see, close by the coast, the great city where the lights were winking like a hundred stars, hear the music and the noise and bustle of carts and people, see the towers and spires of the many churches and listen to the ringing of their bells. And just because she could not go up into the city, that was what she longed to do more than anything else.

Oh, how intently her youngest sister listened to her! And whenever, after that, she stood in the evening by the open window, looking up through the dark-blue sea, she thought of the great city with all its noise and bustle, and then she seemed to hear the church-bells ringing down to her.

The next year the second sister was allowed to rise up through the water and swim wherever she wished. She popped her head up just as the sun was going down, and that sight she found the most beautiful. The whole sky had looked like gold, she said, and the clouds—she just could not describe their loveliness! Red and violet, they had sailed away over her head, but far swifter than they, flew a flock of wild swans, like a long white ribbon, away over the water to where the sun stood; she swam off towards it, but it sank, and the rosy gleam disappeared from the surface of the sea and the clouds.

The year after, the third sister went up above—she was the most venturesome of them all, and so she swam up a broad river which flowed into the sea. She saw lovely green hills with vineyards upon them, and castles and farms peeping out from the magnificent forest. She heard all the birds singing, and the sun shone so warmly, she often had to duck under the water to cool her burning face. In a little bay she came across a crowd of little children running about quite naked and splashing in the water. She wanted to play with them, but they ran off in fright. Then a little black animal came along—it was a dog, but she had never seen a dog before. It barked at her so terrifyingly that she became alarmed and made off for the open sea, but she could never forget the magnificent forest, the green hills, and the pretty children who could swim in the water, even though they had no fishes' tails.

The fourth sister was not so venturesome. She stayed out in the middle of the wild sea, and told them that was quite the loveliest place; you could see for many miles all round you, and the sky stood overhead like a huge glass bell. She had seen ships, but far away, looking like gulls; playful dolphins had turned somersaults, and great whales had spouted water up from their nostrils so that it looked as if there were hundreds of fountains round about.

Now came the turn of the fifth sister. Her birthday happened to fall in winter, and so she saw what the others, during their first visits, had not seen. The sea looked quite green, and great icebergs were swimming round about, each one looking like a pearl, she said, and yet they were far larger than the church towers built by men. They appeared in the strangest shapes, glittering like diamonds. She had seated herself on one of the largest, and, in terror, ships sailed wide of the ice where she sat, her long hair streaming in the blast. Late in the evening the sky became overcast with cloud, and it thundered and lightened, while the dark sea lifted the great ice-floes high up and made them shine in the strong flashes of lightning. On all ships, sails were taken in, and there was alarm and fear, but she sat peacefully on her floating iceberg, and watched the blue flashes of lightning strike zig-zagging down upon the shining sea.

The first time the sisters, each in her turn, came to the surface of the sea, they were always delighted with the new and beautiful sights they saw, but now that they were grown girls and had leave to go up there whenever they wished, they were quite indifferent about it and longed for home again, and at the end of a month they said that the bottom of the sea was the most beautiful place of all, the place where they felt most comfortable and most at home.

Many an evening the five sisters would take one another by the arm and swim in a row up to the surface of the water. They had lovely voices, more beautiful than any human voice, and when the wind was blowing up to a gale and they thought ships might be lost, they would swim in front of the ships and sing so beautifully how lovely it was on the bottom of the sea and bid the sailors not to be afraid of coming down there. But the men could not understand their words: they thought it was the gale they heard. Nor did they ever see the beauty down below them, for when their ship sank, the men were drowned, and only as dead men did they ever come to the mer-king's palace.

On those evenings when the sisters rose arm-in-arm like this high up through the sea, their little sister was left behind quite alone, gazing after them, and she looked as though she would weep, but a mermaid has no tears and so she suffers all the more.

'Oh, if only I were fifteen!' she said. 'I know that I shall really grow to love that world above the sea and all the people that live and dwell there!'

And then at last she was fifteen.

'Well, we've got you off our hands now!' said her grandmother, the old queen-dowager. 'Come along now, let me dress you up like your sisters!' And she placed a wreath of white lilies on her hair, but every petal of the flowers was half a pearl. Then the old lady fastened eight large oysters firmly on the princess's tail to show her high rank.

'But it hurts so!' said the little mermaid.

'We must all suffer a little to look smart!' said the old lady.

Oh, she would have loved to shake off all that finery and put down that heavy wreath! The red flowers in her garden suited her much better, but she dared not change them now. 'Good-bye!' she said, and then she rose, lightly and clearly like a bubble, up through the water.

The sun had just gone down as she raised her head up above the water, but the clouds were all still gleaming with rose and gold, and in the middle of the pale-pink sky the evening-star shone clear and lovely, the air was mild and fresh and the sea a glassy calm. A great three-masted ship lay there with only a single sail set, for there was not a breath of wind, and the sailors were sitting about in the rigging and on the spars. There was music and song, and as the evening grew darker, hundreds of coloured lights were lit—it looked as if the flags of all the countries in the world were waving in the air. The little mermaid swam right up to the cabin window, and every time she rose with the swell of the water, she could see through the crystal-clear panes where a large number of well-dressed people were standing, but the most handsome of them was the young prince with the large black eyes. He was certainly not much more than sixteen; it was his birthday, and that was why there was all this celebration. The sailors were dancing on deck, and as the young prince stepped out, over a hundred rockets shot up in the air, shining as brightly as daylight, so that the little mermaid was quite terrified and ducked under the water, but she soon popped her head up again, and it looked just as if all the stars of heaven were falling down on her. She had never seen fireworks before. Great suns were whirling round, magnificent firefish were soaring up into the blue sky, and it was all reflected back from the clear, still sea. On the ship itself, there was so much light, you could see every little rope, let alone the people. Oh, how handsome the young prince was, shaking everyone by the hand, laughing and smiling, while the music rang out into the lovely night!

It grew late, but the little mermaid could not turn her eyes away from the ship and the handsome prince. The coloured lights were put out, no more rockets rose into the air, nor was there any more cannon-fire to be heard either, but deep down in the sea there was a murmuring and a rumbling. Meanwhile, she sat on the water, rocking up and down, so that she could look into the cabin. But the ship gathered speed as one sail was spread after another. Then the waves ran higher, great clouds closed in, and far away there were flashes of lightning. Oh, there was going to be a dreadful storm! And so the sailors took in sail. The great ship sped rolling and swaying through the wild sea, the waves rose like great black mountains about to break over the mast, but the ship dived like a swan down among the high billows and was lifted up again on the piled-up waters. To the little mermaid their speed was pleasant and amusing enough, but the sailors did not think so. The ship creaked and cracked, the stout planks bent under the heavy blows dealt by the sea, the mast snapped in the middle like a reed, and the ship heeled over on her side as the water rushed into her hold. Then the little mermaid saw that they were in danger—she herself had to look out for beams and broken ship's timbers driving on the water. At one moment it was so pitch-black that she could see nothing at all; then when the lightning flashed, it was so bright again she could make out all those on board, each man staggering about as best he could. She was searching especially for the young prince, and as the ship broke up, she saw him sink down in the deep sea. For the moment she was overjoyed, for now he was coming down to her, but then she remembered that men could not live in the water, and that he could go down to her father's palace only as a dead man. No, die he must not! And so she swam off among the beams and planks driving on the sea, completely forgetting that they could have crushed her, dived deep under the water and rose again high up among the waves, and then at last she reached the young prince who could scarcely swim any farther in that stormy sea. His arms and legs were beginning to tire, his beautiful eyes had shut, and he would have been left to die had the little mermaid not come to him. She held his head up above the water and let the waves drive her with him wherever they would.

When morning came the storm had passed, but of the ship there was not a stick to be seen. The sun rose out of the water so red and bright, the prince's cheeks seemed to gain life from it, but his eyes remained shut. The mermaid kissed his beautiful high forehead and stroked his wet hair back; she thought he looked like the marble statue down in her little garden; she kissed him again and wished that he might yet live.

Then she saw dry land in front of her, high blue mountains with snow shining white upon their summits as if it were swans that lay there; down by the coast were lovely green woods and before them lay a church or an abbey—she did not rightly know what, but at least a building. Oranges and lemons grew there in the garden, and in front of the gate tall palm-trees stood. The sea made a little bay there, dead still but very deep right up to the rocks where the fine white sand was washed up. She swam over to that point with the

handsome prince, laid him upon the sand and took particular care that his head lay well up in the warmth of the sun.

Then the bells rang in the great white building, and many
young girls came out through the garden. The little mermaid
then swam farther out behind some high rocks that jutted
out of the water, covered her hair and her breast in sea-foam
so that no one could see her little face, and then watched to
see who would come down to the poor prince.

It was not long before a young girl came down there; she
seemed quite frightened, but only for a moment, and then
she went and fetched some more people. And the mermaid
saw the prince revive and smile at everybody round about
him, but he did not turn and smile at her, for of course he did
not even know that she had rescued him. She felt so sad that
when he was carried into the great building, she dived
sorrowfully down into the water and made her way home to
her father's palace.

She had always been quiet and thoughtful, but now she
became even more so. Her sisters asked her what she had
seen the first time she had been above the sea, but she told
them nothing.

Many an evening and morning she rose up where she had
left the prince. She watched the fruit in the garden ripen and
she saw it gathered in; she saw the snow melt on the high
mountains, but the prince she did not see, and so, every time,
she turned home even more sorrowful than before. Her one
comfort was to sit in her little garden and throw her arms
round the beautiful marble statue that looked so like the
prince. But she no longer looked after her flowers, and they
grew, as if in a wilderness, out over the paths, and twisted
their long stems and their leaves in among the branches of
the trees until it was quite dark there.

At last she could bear it no longer, and told one of her
sisters all about it, and so all the others got to know about it,
too, right away, but no one else beyond them and a couple of
other mermaids, who, of course, told no one, except their

closest friends. One of them knew who the prince was—she, too, had seen the party on board the ship, and she knew where he was from and where his kingdom lay.

'Come, little sister!' said the other princesses, and with their arms about each others' shoulders they rose up out of the sea in a long row in front of the spot where they knew the prince's palace lay.

It was built of a gleaming pale-yellow stone, with great marble steps, some leading right down into the sea. Magnificent gilt domes rose above the roof, and between the pillars which surrounded the whole building stood marble statues that looked as if they were alive. Through the clear glass of the tall windows, you could see into the splendid halls where costly silk curtains and tapestries were hung and all the walls were adorned with great paintings that it was a real joy to look upon. In the middle of the largest hall splashed a great fountain, its jets of water mounting high up towards the glass dome in the ceiling through which the sun shone upon the water and the lovely plants that grew in the great basin.

Now she knew where he lived, and there she came to the surface of the sea many an evening and night. She would swim much closer to the shore than any of the others had ever dared to do, yes, she would go right up the narrow canal below the magnificent marble balcony which cast its long shadow out over the water. There she would sit and look at the young prince who thought he was quite alone in the clear moonlight.

She saw him many an evening sailing to music in the fine boat where the flags were flying. She would peep out from among the green rushes, and if the wind caught her long silver-white veil and anyone saw it, he thought it was a swan spreading its wings.

Many a night, when the fishermen lay with their lights upon the water, she heard them speaking so well of the young prince, and that made her glad she had saved his life

70

when he had driven half-dead upon the waves, and she would remember how closely his head had rested upon her breast and how deeply she had kissed him—and he knew nothing at all about it, and could not even dream of her.

She came to like people more and more, more and more she wished she could go up among them; their world seemed to her far greater than her own, for they could sail away over the sea in ships and climb upon the high mountains high above the clouds, and the lands they owned stretched farther, with forests and fields, than her eye could reach. There was so much she wanted to know, but her sisters were not able to give answers to everything, and so she questioned their old grandmother, who knew the upper world, as she very rightly called the lands above the sea, very well indeed.

'If they are not drowned, can people live for ever?' asked the little mermaid. 'Don't they die, as we do down here in the sea?'

'Yes, of course, they do!' the old lady said. 'They have to die, too, and their lifetime is even shorter than ours. We can live until we are three hundred years old, but then when our life is finished here, we are only foam upon the water, we do not even have a grave down here among our dear ones. We have no immortal soul, we never again come to life, we are like the green rushes—once they are cut through, they cannot grow green again! But people on the other hand have a soul which lives for ever, which goes on living after the body has become dust; it rises up through the bright sky, up to all the shining stars! Just as we dive up out of the sea to look at the lands where people live, so they dive up to lovely unknown places that we shall never come to see.'

'Why were we not given an immortal soul?' asked the little mermaid sadly. 'I would give all the hundreds of years I have to live to be a human girl for just one day and then to receive my part in the Kingdom of Heaven!'

'You must not go thinking about such things,' said the old

71

lady. 'We have a much happier and better time than the people up there do.'

'Then I must die and float like foam upon the sea, and not hear the music of the waves or see the lovely flowers and the red sun. Can't I do anything at all to gain an eternal soul?'

'No,' said the old lady. 'Only if a human man should fall in love with you so deeply that you would be more to him than father and mother; only if he should cling to you with all his thought and all his love, and let the priest place his right hand in yours with a vow of faith here and for all eternity: then his soul would overflow into your body, and you, too, would have your share in human fortune. He would give you a soul and yet keep his own. But that can never happen! What is especially delightful here in the sea—your fishtail—they find ugly up there on earth: they won't think any the better of you for it—there you must have two clumsy props they call legs to be thought beautiful.'

Then the little mermaid sighed and looked sadly at her fishtail.

'Let us be joyful!' said the old lady. 'We'll skip and dance through the three hundred years we have to live. It's a good long time really, and afterwards we shall be able the more contentedly to rest. We'll hold a court ball this evening!'

There was a splendour such as you would never see on earth. The walls and the ceiling of the great ballroom were of thick but clear glass. Several hundred huge mussel-shells, rose-red and grass-green, stood in rows on either side with burning blue flames in them, lighting up the whole room and shining out through the walls, so that the sea outside was quite brightly lit. You could see the countless fish, great and small, that came swimming towards the glass walls: on some shone scales of purple-red, on others they seemed to be of silver and gold. Through the middle of the hall flowed a broad running current, and on this the mermen and women danced to their own lovely singing. Such beautiful voices are not to be found among the people on earth. The little mermaid sang the most beautifully of them all, and they all applauded her. And, for the moment, she felt joy in her heart, for she knew she had the most beautiful voice of all who lived upon the earth or in the sea. But she very soon found herself thinking once more of the world above her: she could not forget the handsome prince and her own sorrow at not having, like him, an immortal soul. And so she stole out of her father's palace, and while all was song and merriment inside, she sat sadly in her little garden. Then she heard the sound of a French horn ring down through the water, and she thought, 'He must be sailing up there now, him I hold dearer than father and mother, him my thoughts cling to, and into whose hands I would put my life's happiness. I would risk everything to win him and an immortal soul! While my sisters are dancing there in my father's palace, I will go and see the sea-witch—I have always been so frightened of her, but perhaps she can advise and help me.'

Then the little mermaid went out of her garden, away towards the roaring whirlpools behind which the witch dwelt. She had never been that way before. No flowers grew there, no sea-grass—only the naked grey sand of the sea-bed stretched away towards the whirlpools that whirled round like roaring mill-wheels and bore away everything that came within their grasp down with them into their depths. Right through the midst of these tremendous whirlpools she had to go to reach the sea-witch's district, and there, for a long stretch, there was no other way but over the warm bubbling mud the witch called her peat-bog. Beyond this lay her house in the middle of a lonely forest. The trees and bushes were all polyps, half beast, half plant, that looked like hundred-headed snakes growing out of the earth. All the branches were long slimy arms with fingers like supple worms, and limb by limb they were in constant motion from their roots to their topmost tips. Everything they could seize upon in the ocean they twined themselves fast about and never more let go. The little mermaid grew quite terrified as she stood just outside the forest. Her heart beat with fear and she very nearly turned back, but then she thought of the prince and the soul of man, and so she found courage. She bound her long streaming hair firmly about her head so that the polyps should not be able to seize her by it, and placed both her hands together over her breast, and in this way she flew through the water like a fish in among the grisly polyps that stretched their supple arms and fingers after her. She saw that each of them had something it had seized, and hundreds of small arms held it like strong bands of iron. Men who had perished at sea and had sunk deep down below peeped out like white skeletons from the polyps' arms. Ships' rudders and sea-chests they held in their grip, and the skeletons of land animals and a little mermaid they had seized and strangled—that made her more afraid than almost anything else.

Then she came to a great slimy opening in the forest where big fat water-snakes were frisking about and showing their ugly yellow-white bellies. In the middle of the opening, a house had been built from the bones of shipwrecked men. There the sea-witch sat letting a toad eat out of her mouth, just as people will let a little canary eat sugar. The ugly fat water-snakes she called her little chickens, and she let them roll about on her great spongy breasts.

'I know well enough what it is you want,' said the sea-witch. 'You're acting like a fool! However, you shall have your way, for it will bring you misfortune, my lovely princess. You'd like to get rid of your fishtail and have two stumps in place of it so that you can walk like a human. Then the young prince will fall in love with you, and you'll be able to get him and an immortal soul as well.' At this the witch laughed so loudly and unpleasantly that the toad and the snakes fell rolling on to the ground. 'You've come just at the right time,' said the witch. 'If it had been after sunrise tomorrow morning, I could not have helped you until another year had passed. I shall make you a drink: before the

74

sun rises, you must swim to land with it, sit yourself down upon the shore, and drink it. Then your tail will divide and shrink into what men call a lovely pair of legs. But it will hurt: it will be like a sharp sword going through you. Everybody who sees you will say you are the loveliest human child they have ever seen! You will keep your gliding motion: no dancer will be able to glide along like you, but every step you take will be like treading on a sharp knife that cuts you and makes your blood flow. If you are willing to suffer all this, then I will help you.'

'Yes,' the little mermaid said in a trembling voice, and she thought of the prince and of winning an immortal soul.

'But remember,' said the witch, 'once you have taken human form, you can never become a mermaid again! You will never be able to go down through the water to see your sisters or your father's palace. And if you do not win the prince's love so that he will forget father and mother for your sake, cling to you with all his mind and let the priest place your hands in one another's so that you become man and wife, you will not get an immortal soul. The very first morning after he weds another, your heart will break, and you will become foam upon the water.'

'I am willing,' said the little mermaid, and she was as pale as death.

'But you must pay me, too,' said the witch. 'And it is no small thing I am asking for. You have the loveliest voice of all down here upon the bottom of the sea, and no doubt you think you will be able to bewitch him with it, but that voice you must give to me. The best thing you possess will I have in return for my precious drink. For I must give you my own blood in it to make the drink sharp like a two-edged sword.'

'But if you take my voice,' said the little mermaid, 'what shall I have left?'

'Your lovely figure,' said the witch, 'your gliding walk and your speaking eyes—with them you will be able to charm a man's heart, never you fear! Well, have you lost your courage? Put out your little tongue so that I can cut it off in payment, and you shall have your powerful drink.'

'It shall be done,' said the little mermaid, and the witch put her cauldron on to boil the magic potion. 'Cleanliness is a good thing,' she said, and scoured the cauldron out with the snakes which she had tied into a knot. Then she scratched herself on the breast and let her black blood drip into it. The steam formed shapes so strange and wonderful that they would have filled you with fear and dread. Every second the witch put something new into the cauldron, and when the brew was properly boiling it sounded like a crocodile weeping. At last the drink was ready—and it looked like the clearest water!

'There it is!' said the witch, and cut out the tongue of the little mermaid, who was now dumb and could neither sing nor speak.

'If the polyps should catch hold of you as you go back through my woods,' said the witch, 'you have only to throw one single drop of this potion over them and their arms and fingers will burst into a thousand pieces.' But the little mermaid had no need to do it, for the polyps drew back in fear of her when they saw the shining drink glistening in her hand like a sparkling star. And so she soon returned through the forest, the bog, and the roaring whirlpools.

She could see her father's palace: the lights were out in the great ballroom; no doubt they were all asleep in there, but yet she dared not go and look at them—she was dumb now and had made up her mind to go right away from them for ever. It seemed as if her heart would break with sorrow. She slipped into the garden, took one flower from each of her sisters' flower-beds, blew a thousand kisses towards the palace, and rose up through the dark-blue sea.

The sun had not yet risen when she saw the prince's palace and clambered on to the stately marble steps. The moon was shining lovely and clear. The little mermaid drank the sharp burning drink, and it felt as if a two-edged sword went through her delicate body. She fainted with the pain of it and lay as if she were dead. When the sun shone over the sea, she woke up and she felt a stinging pain, but right in front of her stood the handsome young prince, his coal-black eyes fastened upon her so that she dropped her own and saw that her fishtail had gone and that she had the nicest little white legs any young girl could have, but she was quite naked and so she wrapped her long thick hair about her. The prince asked who she was and how she had come there, and she looked gently yet sadly at him with her dark-blue eyes, for speak she could not. Then he took her by the hand and led her into the palace. Every step she took was, as the witch had told her it would be, like treading on pointed tools and sharp knives, but she bore it all willingly: the prince's hand on hers, she stepped as light as a bubble, and he and everyone there marvelled at her graceful gliding walk.

She was given costly clothes of silk and muslin to wear, and she was the most beautiful girl in the palace, but she was dumb and could neither sing nor talk. Beautiful slave-girls, clad in silk and gold, appeared and sang for the prince and his royal parents: one sang more beautifully than all the rest, and the prince clapped his hands and smiled at her. And then the little mermaid grew sad for she knew that she herself had sung far more beautifully. She thought, 'Oh, if only he could know that, to be near him, I have given my voice away for ever!'

Then the slave-girls danced graceful swaying dances to the noblest music, and the little mermaid raised her beautiful white arms, rose upon the tips of her toes and glided away over the floor, dancing as no one had ever danced before. With every movement her loveliness grew even more apparent, and her eyes spoke more deeply to the heart than the slave-girls' song.

Everyone was enraptured by it, and especially the prince, who called her his little foundling, and she danced on and on, although every time her foot touched the ground it was like treading on sharp knives. The prince said that she should remain with him always, and she was given permission to sleep outside his door on a velvet cushion.

He had a man's suit made for her so that she could go riding with him. They rode through the scented forest where the green branches brushed her shoulders and the little birds sang among the fresh leaves. She went climbing with the prince on the high mountains, and although her delicate feet bled so that all could see, she laughed at it and went with him until they saw the clouds sailing down below them like a flock of birds making their way to distant lands.

At home in the prince's palace, when the others were asleep at night, she would go out on to the broad marble steps, and standing in the cold sea-water would cool her burning feet, and then she would think of those down below in the deep.

One night her sisters came arm-in-arm, singing so sorrowfully as they swam over the water, and she waved to them and they recognized her and told her how sad she had made them all. After that they came to see her every night, and one night she saw, far out, her old grandmother, who had not been to the surface of the sea for many years, and the mer-king with his crown upon his head. They stretched their hands out towards her, but they dared not come as near the land as her sisters.

Day by day she grew dearer to the prince: he was fond of her as one might be fond of a dear good child, but it never occurred to him to make her his queen, and his wife she had to be, or else she would not find an immortal soul but on the morning of his wedding would become foam upon the sea.

'You are fonder of me than of all the others?' the little mermaid's eyes seemed to say when he took her in his arms and kissed her fair brow.

'Yes, of course you are dearest to me,' said the prince, 'because you have the best heart of all of them. You are the one I am most fond of, and you are like a young girl I once saw but shall never find again, no doubt. I was on board a ship which was wrecked, and the waves drove me to land near a holy temple served by several young maidens. The youngest of them found me there by the seashore and saved my life. I saw her only twice: she is the only one I could love in all the world, but you are like her—you have almost taken the place of her picture in my soul. She belongs to the holy temple, and so good fortune has sent me you—never will we part from one another!'

'Ah, he does not know that I saved his life!' thought the little mermaid. 'I bore him over the sea away to the woods where the temple stands, I hid behind the foam and watched to see if anyone would come. I saw the pretty maiden he loves more than me.' And the mermaid sighed deeply—weep she could not. 'The maiden belongs to the holy temple, he said. She will never come out into the world, they will not meet again, and I am here with him, I see him every day—I will look after him and love him and give up my life to him!'

But now everybody was saying that the prince was to marry and have the neighbouring king's lovely daughter, and it was for that purpose he was fitting out such a splendid ship. 'The prince is travelling to see the neighbouring king's land. That's what they're saying, of course, but it's really to see the neighbouring king's daughter. And he's taking a great following with him.' But the little mermaid shook her head and laughed; she knew the prince's thoughts much better than the rest of them. 'I have to make a journey,' he had told her. 'I have to go and see this lovely princess. It's my parents' wish—but they won't force me to bring her home as my bride: I cannot love her. She will not look like the lovely maiden in the temple, as you do. If I should ever choose a bride, then it would sooner be you, my dumb foundling with the speaking eyes.' And he kissed her red mouth, played with her long hair, and laid his head against her heart so that it dreamed of human happiness and an immortal soul.

'You do not seem a bit afraid of the sea, my dumb child,' he said, as they stood on the splendid ship which was to carry him to the land of the neighbouring king. And he told her of storm and calm, of rare fish in the deep and of what divers had seen. And she smiled at what he told her, for, of course, she knew more than anyone about the bottom of the sea.

In the moon-clear night when they were all asleep except the helmsman standing by the wheel, she sat by the ship's rail and gazed down through the clear water, and she thought she could see her father's palace: on the very top of it her old grandmother was standing with her silver crown upon her head, and gazing up through the swift currents towards the ship's keel. Then her sisters came up to the surface; they gazed sorrowfully at her and wrung their white hands. She waved to them and smiled and wanted to tell them that all was going well and happily with her, but the ship's boy drew near her and her sisters dived below so that he was left believing that the whiteness he had seen was only foam upon the water.

The next morning the ship sailed into the harbour of the neighbouring king's fine city. All the church bells were ringing, and trumpets blared from the high towers while the soldiers stood to attention with waving flags and glinting bayonets. Every day there was merry-making: balls and parties followed one upon another. But the princess was not there yet: she was being brought up far away in a holy temple, they said, where she was learning all the royal virtues. At last she arrived.

The little mermaid stood there eager to see her beauty, and, she had to admit it, a more graceful figure she had never seen. Her skin was so delicate and pure, and behind her long dark lashes smiled a pair of steadfast dark-blue eyes.

'It is you!' said the prince. 'You who saved me when I lay like a corpse on the shore!' And he held his blushing bride tightly in his arms. 'Oh, I am all too happy!' he said to the little mermaid. 'The best thing I could ever hope for but never dared to, has happened to me. You will be glad I am happy, for you are fonder of me than the rest of them.' And the little mermaid kissed his hand, and she thought she felt her heart breaking. His wedding morning would be her death and change her into foam upon the sea.

All the church bells rang, the heralds rode through the streets and proclaimed the betrothal. Upon all the altars burned sweet-scented oil in precious silver lamps. The priests swung their censers, and bride and bridegroom took one another by the hand and received the bishop's blessing. The little mermaid stood in silk and gold holding the bride's train, but her ears did not hear the festive music, her eyes did not see the holy ceremony, her thoughts were upon the night of her death, upon all she had lost in this world.

That very same evening the bride and bridegroom went on board their ship. The cannon were fired and all the flags were streaming, and amidships a royal tent of gold and purple had been raised and furnished with the loveliest cushions. And there the bridal pair were to sleep in the still cool night.

The sails bellied in the wind, and the ship glided away lightly and smoothly over the clear sea.

When it grew dark many-coloured lamps were lit, and the sailors danced merry dances on deck. The little mermaid could not help thinking of the first time she had dived up to the surface of the sea and had seen the same splendour and rejoicing, and she whirled in the dance with the rest, swerving as a swallow swerves when it is pursued, and they all applauded her in admiration, for never before had she danced so brilliantly. Sharp knives seemed to cut into her delicate feet, but she did not feel it, for the pain in her heart cut yet more sharply. She knew this was the last evening she would ever see him for whom she had forsaken her kindred and her home, given up her lovely voice, and daily suffered unending torment—and he had no idea of it. This was the last night she would breathe the same air as he, or look upon the deep sea and the starry blue sky; an everlasting night without thoughts or dreams awaited her, for she had no soul and could not gain one. And all was rejoicing and merry-making on board until well gone midnight. She laughed and danced with the thought of death in her heart. The prince kissed his lovely bride, and she played with his black hair, and arm-in-arm they went to rest in the splendid tent.

The ship grew still and quiet: only the helmsman stood by the wheel. The little mermaid laid her white arms upon the rail and looked towards the east for the first red of morning—the first ray of the sun, she knew, would kill her. Then she saw her sisters rise up to the surface of the sea: they were pale like her; their beautiful long hair no longer fluttered in the breeze—it had been cut off.

'We have given it to the witch so that she would help us and you would not have to die this night! She has given us a knife—here it is! Can you see how sharp it is? Before the sun rises, you must thrust it into the prince's heart, and as his warm blood splashes on to your feet, they will grow together into a fishtail and you will be a mermaid again, and then you can come down to us in the water and live your three hundred years before you turn into the dead salt foam of the sea. Hurry! Either he or you must die before the sun rises! Our old grandmother is so overcome with sorrow that her white hair has fallen out just as ours fell before the witch's scissors. Kill the prince and come back to us! Hurry!—Do you see that red streak in the sky? In a few minutes the sun will rise, and then you must die!' And they gave a strange deep sigh and sank beneath the waves.

The little mermaid drew aside the purple hangings of the tent, and she saw the lovely bride sleeping with her head upon the prince's breast, and she bent down and kissed his fair forehead. She looked at the sky where the red of morning was shining more and more brightly; she looked at the sharp knife and gazed once again upon the prince who murmured his bride's name in his dreams. She alone was in his thoughts, and the knife quivered in the mermaid's hand—but she flung it far out into the waves which shone red where it fell and looked as if drops of blood had spurted up out of the water. Once more she looked with half-glazed eyes upon the prince, threw herself from the ship down into the sea, and felt her body dissolving into foam.

Then the sun rose out of the sea and its rays fell gently and warmly on the death-cold sea-foam; and the little mermaid had no feeling of death upon her. She looked at the bright sun, and up above her floated hundreds of lovely transparent forms; through them she could see the ship's white sails and the red clouds in the sky. Their voices were music, but so ethereal were their tones that no human ear could hear them, just as no earthly eye could see them. Their own lightness bore them without wings through the air. The little mermaid noticed that she had a body like theirs that rose higher and higher out of the foam.

'To whom am I coming?' she asked, and her voice sounded like those of the other beings, so ethereal that no earthly music could echo its tones.

'To the daughters of the air!' answered the others. 'A mermaid has no immortal soul, and she can never gain one unless she wins the love of a mortal man: her immortality depends upon the power of another. The daughters of the air have no eternal soul either, but they can by good deeds create one for themselves. We are flying to the hot climates, where the warm pestilence-laden atmosphere kills people. There we waft cool breezes, we spread the sweet scent of flowers through the air and send freshness and healing. When we have striven to do what good we can for three hundred years, we gain our immortal soul and are given a share in the eternal happiness of mankind. You, poor little mermaid, have striven with your whole heart for the same thing as we strive for. You have suffered and endured, and raised yourself into the world of the spirits of the air. And now you, too, through your good deeds can create an immortal soul for yourself in three hundred years.'

And the little mermaid lifted her bright arms up towards God's sun, and for the first time she felt tears coming. On the ship there was bustle and life again: she saw the prince with his beautiful bride looking for her—they were staring sadly into the bubbling foam as if they knew she had thrown herself into the waves. Unseen, she kissed the bride's brow, smiled upon the prince, and with the other children of the air she soared up upon a rose-red cloud sailing through the air.

'In three hundred years we shall float just like this into the Kingdom of God!'

'We might come there even sooner,' whispered one. 'Unseen we float into people's homes where there are children, and for every day we find a good child that makes its parents happy and deserves their love, so God shortens our period of trial. The child does not know when we are flying through the room, and then when we smile upon it in happiness, a year is taken from our three hundred. But if we see a bad, naughty child, then we must weep tears of sorrow, and every tear adds a day to the time of our trial.'

The Steadfast Tin-Soldier

THERE were once five and twenty tin-soldiers: they were all brothers, because they had all been born from one old tin spoon. They carried rifles on their shoulders, and they held their heads up and looked straight in front of them. Their uniform was red and blue and very fine indeed. The very first thing they heard in this world, when the lid of the box in which they lay was taken off, was the word 'Tin-soldiers!' shouted by a small boy as he clapped his hands. He had been given them because it was his birthday, and he now set them out on the table. One soldier looked exactly like another—only a single one of them was a little different: he had one leg because he had been cast the last, and there wasn't enough tin left. Yet he stood just as firm on his one leg as the others did on their two, and he's the one the story is really about.

On the table where they were set out were many other toys, but the one that caught the eye most was a fine cardboard castle. Through the little windows you could see right into the rooms. Outside stood little trees round a little mirror which was meant to look like a lake. Wax swans swam on it and were reflected there. It was altogether charming, but the most charming thing about it was undoubtedly a little young lady who stood in the middle of the open castle door: she was cut out of cardboard, too, but she wore a skirt of the finest linen and a narrow blue ribbon draped round her shoulders. This was fastened in the middle by a shining spangle, as big as the whole of her face. This little lady stretched both her arms out, for she was a dancer, and raised one leg so high in the air that the tin-soldier could not find it at all, and believed that she had only one leg like himself.

'She'd be the wife for me!' he thought. 'But she's of very high birth: she lives in a castle, and I have only a box—besides, there are five and twenty of us, and there isn't room for her. But I can try to make her acquaintance.' And so he lay down full-length behind a snuff-box which stood on the table, and from there he had a good view of the fine little lady, who continued to stand on one leg without losing her balance.

When it was late in the evening, all the tin-soldiers were put in their box and the people of the house went to bed. And now the toys began to play—they played at visiting and soldiers and going to fine balls. The tin-soldiers rattled in their box, for they wanted to join in, but they couldn't get the lid off. The nut-crackers jumped head-over-heels, and the slate-pencil made a dreadful noise on the slate: there was such a row that the canary woke up and began to chat with them—in verse, too! There were only two who didn't move

from their places, and they were the tin-soldier and the little dancer: she held herself so upright upon the tip of her toes, with both arms stretched out; he stood just as firmly on his one leg, and his eyes didn't leave her for one second.

Now the clock struck twelve, and with a clatter the lid sprang off the snuff-box: there was no snuff in it, no, but a little black goblin popped up—it was a sort of trick.

'Tin-soldier,' said the goblin, 'will you keep your eyes to yourself!'

But the tin-soldier pretended he hadn't heard.

'Just you wait till tomorrow!' said the goblin.

When morning came and the children were up, the tin-soldier was put over on the window-ledge, and whether it was the goblin or a draught that did it, the window suddenly swung open and out fell the tin-soldier head-first from the third storey. He fell at a dreadful pace, his leg in the air, and ended up upon his head, with his bayonet stuck between the paving-stones.

The maid and the small boy went down at once to look for him, but although they came very near to treading on him, they still couldn't see him. If the tin-soldier had shouted, 'Here I am!' they would have found him right enough, but he considered it wasn't done to cry out when he was in uniform.

Now it began to rain, and the drops fell thicker and thicker until it was a regular shower. When it was over, a couple of street-urchins came by.

'Look!' said one of them. 'There's a tin-soldier! He must go for a sail!'

And so they made a boat from a newspaper, put the tin-soldier in the middle, and away he sailed down the gutter. Both the boys ran along by his side, clapping their hands. Heavens above, what waves there were in the gutter, and what a current there was! It had been a real downpour. The paper boat tossed up and down, and every now and then whirled round so quickly that the tin-soldier became quite giddy. But he remained steadfast, and without moving a muscle, he looked straight in front of him and kept his rifle on his shoulder.

All at once the boat was driven into a long stretch of gutter that was boarded over, and here it was quite as dark as it was in his box at home.

'I wonder where I shall get to now,' he thought. 'Yes, yes, it's all the goblin's fault! Ah, if only the young lady were here in the boat, it might be twice as dark for all I'd care!'

At that very moment there appeared a great water-rat that lived under the gutter-boarding.

'Have you got your passport?' asked the rat. 'This way with your passport!'

But the tin-soldier said nothing, and held his rifle even more tightly. The boat was carried along and the rat followed. Ooh, how it ground its teeth, and shouted to sticks and straws, 'Stop him! Stop him! He's not paid his toll! He's not shown his passport!'

But the current grew stronger and stronger. The tin-soldier could already catch a glimpse of daylight ahead where the boarding left off, but he also heard a roaring which was quite enough to terrify a bolder man than he: just imagine, where the boarding ended, the gutter emptied straight out into a big canal! It would be just as dangerous for him to be swept into that, as it would be for us to go sailing over a great waterfall.

He was already so close to it now that he could not stop himself. The boat swept out. The poor tin-soldier held himself as stiffly as he could—no one should say of him that he as much as blinked an eyelid. The boat spun round three or four times and filled with water right to the brim, so that nothing could stop it from sinking. The tin-soldier stood up to the neck in water, and the boat sank deeper and deeper. The paper became looser and looser, and now the water flowed over the soldier's head. Then he thought of the pretty little dancer whom he would never see again, and these words rang in the tin-soldier's ears:

> Onward, onward, warrior!
> Death's what you must suffer!

Now the paper parted, and the tin-soldier fell through—and he was at once swallowed by a big fish.

Oh, how dark it was inside! It was even worse than it was under the gutter boarding, and it was so cramped, too. But the tin-soldier remained steadfast, and lay full-length with his rifle on his shoulder.

The fish twisted and turned about in the most frightening manner. At last it became quite still, and a flash of lightning seemed to pass right through it. Daylight shone quite brightly, and someone cried out, 'The tin-soldier!' The fish had been caught, taken to market, sold and brought up to the kitchen, where the maid had cut it open with a big knife. She picked the soldier up by the waist with two fingers and took him into the living-room, where they all wanted to see the remarkable man who had travelled about in the inside of a fish; but the tin-soldier's head was not a bit turned by their admiration. They stood him up on the table, and there—well, what wonderful things can happen in the world!—the tin-soldier was in the very same room he had been in before, he saw the very same children, and the toys were standing on the table—there was the fine castle with the lovely little dancer: she was still standing on one leg with the other raised high in the air—she, too, was steadfast. The tin-soldier was deeply moved—he was ready to weep tin tears, but that was hardly the thing to do. He looked at her and she looked at him, but they said nothing.

At that very moment one of the small boys took the tin-soldier and threw him right into the stove—he gave no reason at all for doing it: the goblin in the box must have been to blame.

The tin-soldier stood in a blaze of light and felt terribly

hot, but whether it was really the heat of the fire or the heat of love, he didn't know. His bright colours had completely gone—no one could say whether this had happened on his travels or whether it was the result of his sorrow. He looked at the little dancer, and she looked at him: he felt himself melting, but he still stood steadfast with his rifle on his shoulder. A door opened and the wind caught the dancer: she flew like a sylph right into the stove to the tin-soldier, burst into flame and vanished. Then the soldier melted to a blob of tin, and the next day, when the maid cleared the ashes out, she found him in the shape of a little heart. Nothing remained of the dancer but her spangle, and that was burnt coal-black.

The Emperor's New Clothes

MANY years ago there lived an Emperor who was so uncommonly fond of gay new clothes that he spent all his money on finery. He cared nothing for his soldiers; nor did he care about going to the theatre or riding in the woods, except for one thing—it gave him a chance to show off his new clothes. He had a different suit for every hour of the day, and since he spent so much time changing, instead of saying, as one does of a king, 'He is in his council chamber,' they said, 'The Emperor is in his wardrobe.'

Life was very entertaining in the big city where he had his court. Strangers arrived every day, and one day there appeared two rogues who spread the story that they were weavers who had mastered the art of weaving the most beautiful cloth you can imagine. Not only were the colours and patterns outstandingly lovely, but the clothes made from the cloth had the wonderful property of remaining invisible to anyone who was not fit for his job or who was particularly stupid.

'They would indeed be fine clothes to have!' thought the Emperor. 'With those on, I could find out what men in my kingdom are unfit for the jobs they have. And I should be able to tell the wise men from the fools. Yes, I must have some of that cloth made up for me at once!' And he handed over a large sum of money to the two rogues to enable them to begin the work.

So they set their two looms up and looked as if they were hard at work, but there was nothing at all on the looms. They boldly demanded the finest silk and gold thread, which they put in their haversacks, and went on pretending to work at the empty looms until far into the night.

'I should like to know how far they've got with my cloth,' thought the Emperor. But when he remembered that no one who was stupid or unfit for his job could see it, he felt somewhat hesitant about going to see for himself. Now, of course, he was quite certain that, as far as he was concerned, there were no grounds for fear, but nevertheless he felt he would rather send someone else first to see how they were getting on. Everybody in the city knew what wonderful powers the cloth had, and they were all very anxious to see how incompetent and stupid their neighbours were.

'I'll send my honest old minister to the weavers,' thought the Emperor. 'He's the best one to see how the cloth's coming on, for he's sense enough, and no one's fitter for his job than he is.'

So the good-natured old minister entered the room where the two rogues sat pretending to work at the empty looms. 'God help us!' thought the old minister, his eyes wide open. 'I can't see a thing!' But, of course, he was careful not to say so out loud.

Both rogues requested him very politely to take a step nearer, and asked him whether he didn't think the pattern beautiful and the colours charming. As they pointed to the empty loom, the poor old minister stared and stared, but he still could not see anything, for the simple reason that there was nothing to see. 'Heavens above,' he thought, 'surely I am not a stupid person! I must say such an idea has never occurred to me, and it must not occur to anyone else either! Am I really unfit for my job? No, it certainly won't do for me to say I can't see the cloth!'

'Well, you don't say anything,' said the one who was still weaving. 'Don't you like it?'

'Oh, er—it's delightful, quite the finest thing I've ever seen!' said the old minister, peering through his glasses. 'The design and the colours— oh, yes, I shall tell the Emperor they please me immensely.'

'Well, it's very kind of you to say so,' said the two weavers, and they went on to describe the colours and the unusual nature of the pattern. The old minister listened very carefully so that he could say the same thing when he returned to the Emperor. And that is just what he did.

The rogues now demanded a further supply of money, silk, and gold, which they said they must have for their work. But they put it all in their own pockets, and not a single thread ever appeared on the loom. However, they continued, as before, to weave away at the empty loom.

Soon afterwards the Emperor sent another unsuspecting official to see how the weaving was going on and whether the cloth would soon be finished. The same thing happened to him that happened to the minister: he stared and stared, but as there was nothing but the empty loom, he could not see a thing.

'Yes, it's a lovely piece of stuff, isn't it?' said the two

rogues. And they showed him the cloth, and explained the charming pattern that was not there.

'Stupid I most certainly am not!' thought the official. 'Then the answer must be that I am not fit for my job, I suppose. That would be a very odd thing, and I really can't believe it. I shall have to see that no one else suspects it.' And then he praised the cloth he could not see, and assured them how happy he was with the beautiful colours and the charming pattern. 'Yes,' he told the Emperor, 'it's quite the finest thing I've ever seen!'

The story of the magnificent cloth was now on everybody's lips.

And now the Emperor wanted to see it himself while it was still on the loom.

With a large number of carefully chosen courtiers—among them the two good old men who had been there before—he paid a visit to the two crafty rogues, who were weaving away with all their might, but with neither weft nor warp.

'Isn't it really magnificent?' asked the two officials. 'Will Your Majesty be pleased to examine it? What a pattern! What colours!' And they pointed to the empty loom, fully believing that the others could undoubtedly see the cloth.

'What's this!' thought the Emperor. 'I don't see a thing. This is really awful! Am I stupid? Am I not fit to be Emperor? That would be the most shocking thing that could happen to me!—Oh, it's very beautiful,' he said aloud. 'It has my very highest approval.' He nodded in a satisfied manner, and looked at the empty loom: on no account would he tell anyone that he could not see anything. All the courtiers who had come with him stared and stared, but none of them could make out any more than the others. But they all repeated after the Emperor, 'Oh, it's very beautiful!' And they advised him to have a suit made of the wonderful new cloth so that he could wear it for the first time for the great procession that had been arranged. 'It's magnificent! Delightful, excellent!' was repeated from mouth to mouth, and they all appeared to be deeply impressed and delighted with it. The Emperor gave each of the rogues an Order of Knighthood to hang in his buttonhole and the title of Knight of the Loom.

The two rogues sat up the whole night before the morning when the procession was to take place, and they had sixteen candles burning. Everyone could see that they had a job on to get the Emperor's new clothes ready in time. They pretended to take the cloth off the loom, they cut out large pieces of air with their big tailor's scissors, they sewed away with needles that had no thread in them, and at last they said, 'Look, the clothes are ready!'

The Emperor with the most distinguished of his gentlemen came to see for himself, and the rogues both held one arm up as if they were holding something, and they said, 'Look, here are the trousers. Here's the jacket. This is the cap.' And so on and so on. 'They are as light as gossamer. You'd think you'd nothing on your body, and that, of course, is the whole point of it.'

'Yes,' said all the gentlemen, but they couldn't see anything, because there was nothing there.

'Will Your Imperial Majesty most graciously be pleased to take your clothes off?' said the rogues. 'Then we shall put the new ones on Your Majesty over here in front of the big mirror.'

The Emperor laid aside all his clothes, and the two rogues pretended to hand him his new clothes, one at a time. They put their arms round his waist, and appeared to fasten something that was obviously his train, and the Emperor turned himself round in front of the mirror.

'My, how well it suits His Majesty! What a perfect fit!' they all said. 'What a pattern! What colours! It must be worth a fortune!'

'The canopy which is to be borne over Your Majesty in

the procession is waiting outside,' said the Chief Master of Ceremonies.

'Right,' said the Emperor, 'I'm quite ready. Doesn't it fit well?' And he turned round once more in front of the mirror and pretended to take a good look at his fine suit.

The gentlemen of the chamber, whose job it was to bear the train, fumbled on the floor with their hands as if they were picking it up, and then they held their hands up in the air. They dared not let anyone notice that they couldn't see anything.

And so the Emperor walked in the procession under his fine canopy, and everybody in the streets and at their windows said, 'My, look at the Emperor's new clothes! There's never been anything like them! Look at the beautiful train he has to his coat! Doesn't it hang marvellously?' No one would let anyone else see that he couldn't see anything, for if he did, they would have thought that he was not fit for his job, or else that he was very stupid. None of the Emperor's clothes had ever had such a success before.

'But, Daddy, he's got nothing on!' piped up a small child.

'Heavens, listen to the voice of innocence!' said his father. And what the child had said was whispered from one to another.

'He's nothing on! A little child said so. He's nothing on!'

At last, everybody who was there was shouting, 'He's nothing on!' And it gradually dawned upon the Emperor that they were probably right. But he thought to himself, 'I must carry on, or I shall ruin the procession.' And so he held himself up even more proudly than before, and the gentlemen of the chamber walked along carrying a train that was most definitely not there.

The Wild Swans

FAR away from here, where the swallows fly when we have winter, there lived a king who had eleven sons and one daughter, Elisa. The eleven brothers—they were princes, of course—went to school with stars upon their breasts and swords by their sides: they wrote upon golden slates with diamond slate-pencils, and said their lessons off by heart just as well as they could read them from a book—you could see at once that they were princes. Their sister Elisa sat on a little stool made of looking-glass, and had a picture-book which had cost half a kingdom.

Oh, yes, the children had a very good time, but it was not always to be so.

Their father, who was king over the whole land, married a wicked queen, who was not at all kind to the poor children. This was made clear to them from the very first day: a great celebration was held throughout the palace, and so the children played at visiting, but instead of getting all the cakes and baked apples that were left over, as they used to do, they were given nothing but sand in a tea-cup, and told to make believe with that.

The week after, the queen sent the little sister, Elisa, away to the country to live with a peasant family, and it was not long before she turned the king so much against the poor princes that he would have nothing more to do with them.

'Fly out into the world and shift for yourselves!' said the wicked queen. 'Fly away like great birds without voices!' She had not the power, however, to do them all the evil she would, and they became eleven beautiful wild swans. With a strange harsh cry, they flew out of the palace windows away over the park and the woods.

It was still quite early in the morning when they passed by the peasant's cottage where their sister Elisa lay asleep; they hovered over the roof, craned their long necks this way and that, and beat their wings, but no one heard them or saw them. Off they had to go again, high up towards the clouds, far away into the wide world, where they flew over a great dark forest which stretched right down to the sea-shore.

Poor little Elisa stood in the peasant's living-room and played with a green leaf—she had no other toys. She stuck a hole in the leaf, peeped through it up at the sun, and it was just as if she saw her brothers' clear eyes; and every time the warm sun-beams shone upon her cheeks, she thought of her brothers' kisses.

One day went by just like another. Whenever the wind blew through the great rose-hedge outside the cottage, it would whisper to the roses, 'Who could be more beautiful than you?' But the roses would shake their heads and say, 'Elisa!' And whenever the old woman sat in the doorway on

Sundays reading her hymn-book, the wind would turn the pages over and say to the book, 'Who could be more pious than you?' 'Elisa!' the hymn-book would answer. And what the roses and the hymn-book said was the pure truth.

When she was fifteen she had to return home, and when the queen saw how beautiful she was she grew angry and spiteful. She would willingly have changed her into a wild swan like her brothers, but she dared not do that right away, because the king would want to see his daughter.

In the early morning the queen went into the bathroom, which was built of marble and decorated with soft cushions, and beautiful rugs, and she took three toads with her, kissed them, and said to the first, 'When Elisa gets into the bath, sit upon her head and make her dull and lazy like you! You sit on her forehead,' she said to the second, 'and make her ugly like you, so that her father will not recognize her! Rest upon her heart,' she whispered to the third, 'and give her evil thoughts to torment her!' So she put the toads in the clear water, which immediately took on a greenish tinge, called Elisa, undressed her, and made her step into the water. As soon as she lay down, the first toad settled in her hair, the second on her forehead, and the third upon her breast, but Elisa did not seem to notice anything at all. When she stood up there were three red poppies floating on the water—if the creatures had not been poisonous and kissed by the witch, they would have changed into red roses. They became flowers, even so, just from resting upon her head and her heart: she was too pious and innocent for witchcraft to have any power over her.

When the wicked queen saw that, she rubbed her with walnut juice so that she became quite dark brown, she smeared her lovely face with an evil-smelling ointment and matted her beautiful hair: it was impossible to recognize the lovely Elisa any longer.

And so when her father saw her he was quite alarmed and said it was not his daughter. No one else would own her, either, except the watch-dog and the swallows, and they were only humble beasts, and there was nothing they could say.

Then poor Elisa wept and thought of her eleven brothers who were all far away. Sadly she stole out of the palace, and all day long she wandered over field and bog until she entered the great forest. She had no idea where she would go, but she felt so sad and longed so much for her brothers, who had been driven out into the world like herself, that she made up her mind to try to find them.

She had been in the forest only a short while when night fell: she had left path and roadway far behind. Then she lay down on the soft moss, said her evening prayers, and leaned her head against the stump of a tree. It was all so still, the air was mild, and round about her in the grass and on the moss hundreds of glow-worms shone like a green fire, and when she touched one of the branches gently with her hand, the glowing insects fell about her like star-dust.

All night long she dreamed of her brothers: they played again as children, they wrote with their diamond slate-pencils upon their golden slates and they looked at that beautiful picture-book which had cost half a kingdom. But they did not, as they had done before, write nothing but noughts and straight lines upon their slates—no, they wrote of the bold deeds they had done, of all they had seen and been through: and everything in the picture-book came alive—the birds sang, and the people stepped out of the book and talked to Elisa and her brothers, but when she turned over the page, they jumped straight back again so that the pictures should not get muddled.

When she woke the sun was already high: she could not see it, of course, for the high trees spread out their branches thick and fast, but the sunbeams played upon them like a fluttering golden gauze. There was a fresh smell of green leaves, and the birds almost settled on her shoulders. She heard water splashing from many large springs which all fell into a pool with a lovely sandy bottom. True, bushes grew thickly round it, but there was one place where the deer had trodden a large opening, and through this Elisa made her way down to the water. It was so clear that if the trees and bushes had not moved with the wind, she might have thought they were painted on the bottom, so clearly was reflected every leaf, whether it caught the rays of the sun or was quite hidden in the shade.

As soon as she saw her own face she had quite a fright, it was so brown and dreadful, but when she had wetted her little hand and rubbed her eyes and forehead, the white skin shone through again. Then she put aside all her clothes and stepped into the fresh water, and there was not a lovelier king's daughter in the whole world than she was.

When she had once more dressed and plaited her long hair, she went to the gushing spring, drank from the hollow of her hand, and wandered farther into the forest without knowing where she was going. She thought about her brothers and the goodness of God who would surely not forsake her. He made that wild crab apple tree she saw ahead to satisfy the hungry. The branches of the tree were bowed with fruit, and from it she made her midday meal. She set props under its heavy branches, and then entered the darkest part of the forest. It was so still, she heard her own footsteps, she heard every little withered leaf she trod upon. There was not a bird to be seen; not a sunbeam could penetrate the great thick branches of the trees; the tall trunks stood so close together that when she looked straight ahead, it seemed that one great trellis of timber after another shut her in. Yes, here was a loneliness she had never known before.

The night was so dark, too: not a single little glow-worm shone in the moss. Sadly she lay down to sleep: then the branches over her head seemed to part, and she saw Our Lord looking down upon her with gentle eyes, and little angels peeping out over His head and under His arms.

When she woke in the morning she did not know whether she had dreamed it or whether it had really been so.

She had gone some steps forward when she met an old woman with berries in her basket, and the old woman gave her some of them. Elisa asked her if she had not seen eleven princes riding through the forest.

'No,' said the old woman, 'but yesterday I saw eleven swans with golden crowns upon their heads swimming down the stream close by here.'

And she led Elisa a little farther forward to a steep bank. At the bottom of the slope a stream wound its way through the woods. The trees upon its banks stretched their long leafy branches over towards each other, and where their branches were not long enough for them to touch, they had torn their roots loose from the earth and leaned out over the water with the branches intertwined.

Elisa said good-bye to the old woman, and went along down the stream to where it flowed out upon a great open beach.

The beautiful sea lay open before the young girl's eyes, but not a ship appeared upon it, not a boat was to be seen, nothing to carry her farther. She looked at the countless pebbles on the shore: the water had worn them all to a round form—glass, iron, stone, everything that lay washed up on the beach had been moulded by the water, and yet it was far softer than her delicate hand. 'It goes on rolling untiringly, and so all these hard stones are made smooth and even. I will be just as tireless. Thank you for your lesson, clear, rolling waves! Some day, my heart tells me, you will bear me to my dear brothers.'

On the seaweed washed up by the tide lay eleven white swan feathers: she gathered them together like a bunch of flowers. Drops of water lay upon them, but no one could see whether they were dew or tears. It was lonely on the seashore, but she did not notice it, for the sea changed everlastingly—yes, it changed more in a few hours than fresh

inland waters in a whole year. When a great black cloud passed over it, the sea seemed to answer, 'I can look dark, too.' And then the wind would blow strong and the waves turn to white foam. But when the clouds shone brightly tinged with red and the wind slept, then the sea was like a rose-leaf. It was now green, now white, but however peacefully it rested, there was still a slight movement on the shore, the water lifting lightly like the breast of a sleeping child.

When the sun was on the point of setting, Elisa saw eleven white swans with golden crowns upon their heads flying towards the land. They came gliding one behind the other, looking like a long white ribbon. Then Elisa climbed the slope and hid behind a bush. The swans settled near her and beat their great white wings.

As the sun sank below the water, the swan-forms suddenly fell away and there stood eleven handsome princes, Elisa's brothers. She uttered a loud cry, for although they had changed much, she knew they were her brothers. She felt they must be and she sprang into their arms, calling them by their names, and they were very, very happy when they saw and recognized their little sister who was now so grown-up and beautiful. They laughed and they cried, and they soon understood how wicked their step-mother had been to them all.

'We brothers,' said the eldest, 'fly in the likeness of wild swans as long as the sun is in the sky: when it sets we appear as men again. And so, about sunset, we always have to be careful to have a resting-place for our feet, for if we were flying high up in the clouds, we might find ourselves in human form hurtling down into the deep. We do not live here but in a land just as beautiful as this on the farther side of the sea. The way is long from here—we have to cross the great ocean, and there is not a single island on our way where we could spend the night, only a lonely little rock which rises steeply half-way across. It is only just big enough for us all to rest upon it standing side by side, and when the seas are heavy, the spray flies high over us. There we spend the night in our human form. But yet we thank God for it, for without it we could never visit the dear land of our birth, because only on two of the longest days of the year can we make our flight here. Only once a year do we get the chance of visiting the home of our fathers and we dare not stay longer than eleven days, time enough to fly over this great forest to where we can catch a glimpse of the palace where we were born and where father still lives, and see the high tower of the church where mother is buried. Here we have the feeling that even the trees and the bushes are akin to us; here the wild horses race away over the plains just as we used to see them in our childhood; here the charcoal-burner sings the old songs we used to dance to as children—here, in short, is the land of our birth, the land that draws us back, and where we have found you, our dear little sister. We dare not stay here longer than two days now, and then we must away over the sea to a land that is beautiful but not our mother-land. How can we take you with us?—We've neither ship nor boat.'

'How can I free you from the spell?' asked their sister.

And they talked together nearly the whole night through, and dozed but a few hours.

Elisa awoke to the sound of swans' wings swishing over her. Her brothers had been transformed again, and they flew round in great circles and at last far away, but one of them, the youngest, stayed behind. The swan laid his head in her lap and she patted his white wings. They were together the whole day. Towards evening the others came back, and when the sun set they stood in their natural form.

'Tomorrow we fly away from here, and we dare not return for a whole year, but we can't leave you like this! Have you the courage to come with us? My arm is strong enough to carry you through the forest, so should not all our wings together be strong enough to fly with you over the sea?'

'Yes, take me with you!' said Elisa.

They spent the whole night plaiting a net of supple willow bark and tough rushes, and they made it big and strong. Elisa lay down in it, and when the sun rose and the brothers had changed back to wild swans, they seized the net in their beaks and flew high up towards the clouds with their dear sister who lay still sleeping. The sunbeams fell right upon her face, and so one of the swans flew above her head shading her with his broad wings.

They were far from land when Elisa woke: she thought she was still dreaming, so strange did it seem to her to be carried over the sea, high up through the sky. By her side lay a branch of lovely ripe berries and a bundle of tasty roots. The youngest of her brothers had gathered them and put them there for her, and she smiled her thanks to him, for she knew that he was the one who was flying just over her head and shading her with his wings.

They were so high up that the first ship they saw below them looked like a white sea-gull resting upon the water. A cloud lay behind them, a great mountain of a cloud, and Elisa saw her own gigantic shadow and those of the eleven swans flying on it: it was a finer picture than she had ever seen before, but as the sun rose high and the cloud was left farther behind them, the drifting shadow-picture disappeared from view.

All day long they flew on like an arrow whistling through the air, but now they had their sister to carry, it was taking them longer than usual. Bad weather was blowing up and evening was drawing near. Anxiously Elisa watched the sun sinking, and she was still unable to spot the lonely rock in the sea; the swans seemed to her to be beating their wings more strongly. Ah, it was her fault they were not going quickly enough: when the sun set they would change into men, fall into the sea and drown. Then she prayed earnestly in her heart to God, but she still saw no sign of the rock. Black clouds were drawing nearer, strong gusts of wind proclaimed an approaching gale, the clouds stood banked in one great threatening wave and shot forward in a solid leaden mass: lightning followed, flash upon flash.

The sun was now right upon the rim of the sea. Elisa's heart trembled. Suddenly the swans shot downwards so swiftly that she thought she would fall, but just then they straightened out again. The sun was half below the water when she first caught sight of the little rock beneath her: it looked no bigger than the head of a seal sticking up out of the water. The sun was sinking rapidly: it was now only the size of a star. As her foot touched the solid ground the sun went out like the last spark in a piece of burning paper. She saw her brothers standing round her arm in arm, but there was only just enough room for them and her. The sea beat against the rock and flung its spray like a shower of rain over the top of them. The sky was brightly lit by repeated flashes of lightning and the thunder rolled clap upon clap; but the brothers and sister held hands and sang a hymn, and that gave them comfort and courage.

At daybreak the air was clean and still, and as soon as the sun rose the swans flew away from the island with Elisa. The seas were still running high, and when she was high in the air it looked as if the white foam upon the dark-green sea were millions of swans floating on the water.

As the sun rose higher Elisa saw in front of her, half swimming in the air, a land of mountains with shining ice-masses upon the mountain-sides, and in the middle a palace stretching at least a mile in length, with one bold colonnade on top of another. In front waved woods of palm-trees and magnificent flowers as big as mill-wheels. She asked if this were the land she was bound for, but the swans shook their heads, for what she saw was the cloud-palace of Morgan le Fay, ever-changing in its loveliness, and they dared not take any human person there. As Elisa was gazing upon it, woods and palace collapsed and faded, and in their place stood twenty proud churches, all alike, with high towers and pointed windows. She thought she heard the organ play, but it was the sea she heard. Then when she was quite near the churches, they turned into a whole fleet of ships sailing down below her; she looked down, and it was only a sea-mist swirling over the water. Yes, the scene constantly changed before her eyes. Then she saw the real land she was bound for; the loveliest blue mountains rose up with forests of cedar, towns and castles. Long before the sun went down she was sitting on the hill-side in front of a great cave overgrown with delicate green creepers that looked like an embroidered carpet.

'Now we shall see what you'll dream here tonight,' said the youngest brother and showed her her bedchamber.

'If only I might dream how I could set you free!' she said; and her mind was so filled with this thought, she prayed earnestly to God to help her, and even when she had fallen asleep, she went on with her prayer. Then she appeared to be flying high up in the sky to the cloud-palace of Morgan le Fay, and the fairy came to meet her, beautiful and radiant, and yet looking quite like the old woman who had given her the berries in the forest and told her of the swans with the golden crowns.

'Your brothers can be freed,' she said. 'But have you enough courage and endurance? It's true the sea is softer than your delicate hands and yet shapes hard stones, but it doesn't feel the pain your fingers will feel, and it has no heart and doesn't suffer the anxiety and torment you must endure. Do you see this stinging-nettle I am holding in my hand? Many of the same kind grow round about the cave where you are sleeping; only those and the ones that spring up from the graves in the churchyard can be used—mark that! You must pick them—though they will sting you and raise blisters on your skin—tread the nettles with your feet and get a yarn like flax from them, and with that twist a thread and make eleven shirts like coats of mail with long sleeves. Throw these over the eleven wild swans and the spell will be broken. But you must take care to remember that from the moment you begin your task right up to the time it is finished, even though it take years, you must not speak. The first word you utter will pierce your brothers' hearts like a dagger—their lives hang upon your tongue. Remember all I have told you!'

And at the same time she touched Elisa's hand with the nettle; it was like a burning fire. Elisa woke with the pain. It was bright daylight, and close by where she had slept lay a nettle like the one she had seen in her dream. Then she fell upon her knees, thanked Our Lord, and left the cave to begin her task.

She plunged her delicate hands down among the ugly nettles that stung like fire; they burnt great blisters on her hands and arms, but she would suffer gladly if she could free her dear brothers. She trod every nettle with her naked feet, and twisted the green flax into a thread.

When the sun had set, her brothers came back, and they were alarmed to find her so silent; they thought it was a new spell their wicked step-mother had put upon her. But when they saw her hands, they realized what she was doing for their sake, and the youngest brother wept, and where his tears fell she felt no more pain and the burning blisters disappeared.

She spent the night at her task, because she could not rest until she had freed her dear brothers. The whole of the following day, while the swans were away, she sat in solitude, but never had time flown so quickly. One shirt was completely finished, and she began upon the next.

Then a hunting-horn sounded through the mountains; she grew quite frightened as the sound came nearer; she heard the hounds baying, and in terror she fled into the cave, tied together in a bundle the nettles she had gathered and combed, and sat upon it.

At the same moment a great hound sprang out of the bushes, followed immediately by another and still another; they bayed loudly, retreated and sprang forward again. A few minutes later the huntsmen were all standing before the cave, and the most handsome among them was the king of the land. He stepped towards Elisa, for he had never seen a more beautiful girl.

'How do you come to be here, sweet child?' he asked. Elisa shook her head. She dared not speak, for that would cost her brothers' lives and freedom; and she hid her hands under her apron so that the king should not see what she had to suffer.

'Come with me!' he said. 'You cannot stay here. If you are as good as you are beautiful, I will clothe you in silk and velvet and put a golden crown upon your head, and you shall live in my finest castle.' And then he lifted her up upon his horse. She wept and wrung her hands, but the king said, 'I want only your happiness. One day you will thank me for this.' And so, holding her in front of him on his horse, he rode off between the mountains, and the huntsmen followed behind.

When the sun set, the magnificent royal city with its churches and domes lay before them, and the king led her into the palace, where great fountains splashed in high marble halls, their walls and ceilings resplendent with paintings. But she had no eyes for it. She wept and grieved; listlessly, she let the women attire her in royal garments, plait pearls in her hair, and draw fine gloves over her blistered fingers.

When she stood there in all her finery, she was so dazzlingly beautiful that the court bowed even more deeply before her and the king chose her to be his bride, although the archbishop shook his head and whispered that the beautiful girl of the woods was undoubtedly a witch who had blinded their eyes and infatuated the king's heart.

But the king did not listen to him: he commanded the music to strike up, the richest dishes to be brought in, and the loveliest girls to dance about her. She was led through sweet-smelling gardens into magnificent halls; but not a smile passed over her lips or shone from her eyes—sorrow stood there as her inheritance and everlasting possession. Then the king opened a little chamber close by, where she was to sleep. It was decorated with costly green tapestries, and looked quite like the cave where she had been found. On the floor lay the bundle of yarn she had spun from the nettles, and from the ceiling hung the shirt she had finished knitting—one of the huntsmen had brought all this along with him as something of a curiosity.

'In here you can dream you are back in your old home,' said the king. 'And here is the work you were occupied with. It will amuse you now in all your splendour to think on times gone by.'

When Elisa saw what lay so near her heart, a smile played about her mouth and the blood returned to her cheeks. She thought about her brothers' release and kissed the king's hand, and he pressed her to his heart, and commanded all the church bells to proclaim their wedding-feast. The beautiful dumb girl from the woods would be queen of the land.

Then the archbishop whispered evil words in the king's ear, but they made no impression upon his heart. The wedding must stand, the archbishop himself had to place the crown upon her head, and he pressed the narrow ring firmly down over her forehead with such ill-will that it hurt. And yet there lay a heavier ring about her heart in her sorrow for her brothers, so that she did not feel the physical pain. Her mouth was dumb—one single word would cost her brothers their lives—but in her eyes lay a deep love for the good and handsome king, who did all he could to make her happy. Day by day she loved him more and more. Oh, if only she dare open her heart to him and tell him of her sufferings, but silent she had to be, and in silence must she complete her task. And so, at night, she would slip from his side, go to the little room that was decorated to look like the cave, and finished knitting one shirt after another. But when she began upon the seventh she ran out of yarn.

She knew that the nettles she must use grew in the churchyard, but she had to pick them herself—how could she get out to do it?

'Oh, what is the pain in my fingers to the agony my heart suffers!' she thought. 'I must risk it. Our Lord will not take his hand from me.' With fear in her heart, as if it were an evil deed she had to do, she crept down into the garden in the moon-bright night, and made her way down the long avenues, and out into the lonely streets that led to the churchyard. There she saw a coven of ugly witches sitting in a circle on one of the broadest tombstones: they took their rags off as if they were going to bathe, and then with their long skinny fingers they dug down into the fresh graves, pulled out the dead bodies, and fed upon the flesh. Elisa had to pass close by them, and they fastened their evil eyes upon her, but she said her prayers, gathered the stinging-nettles, and carried them home to the palace.

Only one single person had seen her, and that was the archbishop: he was up while the others were asleep. He was now sure he had been right in his opinion that all was not as it should be with the queen. She was a witch and had used her powers to infatuate the king and all the people.

During confession he told the king what he had seen and what he feared, and as the harsh words fell from his tongue, the carven images of the saints shook their heads as if to say, 'It is not so—Elisa is innocent!' But the archbishop took it the wrong way, and thought that they were bearing witness against her, that they were shaking their heads over her sin. Then two heavy tears rolled down the king's cheeks, and he went home with doubt in his heart. He pretended to sleep at night, but no sleep would come to close his eyes in peace. He noticed Elisa get up. Every night she did the same thing, and every time he followed her softly and watched her disappear into her little room.

Day by day his face grew darker. Elisa noticed it, but

could not understand why, though it troubled her and added to the suffering she felt in her heart for her brothers. Her salt tears ran down her royal velvet and purple, and lay there like glistening diamonds; and all who saw her splendour wished they were queen, too. She would soon reach the end of her task, however, for only one shirt still remained to be done, but she had no more yarn left and not a single nettle. And so once more, though this would be the last time, she would have to go to the churchyard and pick a few handfuls. She thought anxiously of the lonely walk and the frightful witches, but her will was as firm as her trust in the Lord.

And so she went, but the king and the archbishop followed her. They saw her disappear by the iron gate that led into the churchyard, and when they drew near, the witches were sitting on the grave-stones just as Elisa had seen them, and the king turned away for he imagined her among them—only that very evening had she laid her head upon his breast.

'I shall let the people judge her!' he said. And the people passed judgement that she should burn in the red flames.

She was borne away from the splendour of the royal halls into a dark damp hole where the wind whistled in through the iron bars of the window. Instead of velvet and silk, they gave her the bundle of nettles she had gathered, to lay her head on, and she had to use the hard stinging shirts she had knitted, as bedding and blankets. But they could have given her nothing she would sooner have had. She turned to her task again, and prayed to God. Outside, the street urchins sang mocking songs about her. Not a soul comforted her with a loving word.

Towards evening came the swish of a swan's wing close by the window-bars—it was the youngest of the brothers: he had found their sister. She sobbed aloud with joy, although she knew that the coming night was probably the last she had to live: but now her work was nearly finished and her brothers were there.

The archbishop came to spend her last hour with her, as he had promised the king he would, but she shook her head, and asked him with looks and gestures to go. That night she had to finish her task, or everything—pain and tears and sleepless nights—everything would have been to no purpose. The archbishop went away speaking hard words against her, but poor Elisa knew she was innocent, and went on with her work.

There were little mice running over the floor, and they dragged the nettles over to her feet, to help a little at any rate, and a thrush settled by the iron window-bars and sang the whole night through as joyfully as it could so that she should not lose heart.

It was still no more than day-break, an hour before sunrise, when the eleven brothers stood at the palace gate and demanded to be taken before the king; but they were told that could not be done—it was still night, the king was asleep and they dared not wake him. They prayed, they threatened; the guard came—yes, even the king himself stepped out to ask what it all meant. At that very moment the sun rose and the brothers were no longer there, but away over the palace flew eleven wild swans.

Everybody streamed out of the city gate to see the witch burnt. A wretched horse dragged the cart in which she sat. She had been given a smock of coarse sackcloth; her lovely long hair hung loose about her beautiful head, and her cheeks were deathly pale. Her lips moved softly while her fingers wove the green yarn—even on her way to death, she did not give up the task she had begun. The ten shirts lay at her feet, and she was knitting away at the eleventh, while the rabble scoffed at her.

'Look at the witch, see the way she mutters ! There's no hymn-book in her hand, you bet! Look at her sitting there with her beastly witchcraft! Tear it away from her! Tear it into a thousand pieces!'

And they all pressed in upon her and would have torn it

apart, but at that moment the eleven white swans came flying towards her. They settled round her on the cart and beat their great wings. The mob drew back in alarm.

'It's a sign from heaven! She must be innocent!' many whispered, but they did not risk saying so aloud.

As the executioner seized her by the hand, she hastily threw the eleven shirts over the swans, and there stood eleven handsome princes, but the youngest had a swan's wing in place of one arm because a sleeve was missing from his shirt which she had not completely finished.

'Now I can speak!' she said. 'I am innocent!'

And the people who saw what had happened bowed before her as if she were a saint, but she sank lifeless in her brothers' arms, overcome by excitement, fear, and pain.

'Yes, she is innocent!' said the eldest brother, and then he told them everything that had happened, and while he was speaking, a perfume like that of a million roses spread about them, for every stick of wood heaped about the stake had taken root and put out branches. There stood a great tall hedge fragrant with red roses. Right at the top was one flower, white and gleaming, which shone like a star. The king plucked it and placed it on Elisa's breast, and then she woke with peace and happiness in her heart.

And all the church bells rang by themselves, and the birds came in great flocks. The way back to the palace became a wedding procession such as no king had yet seen.

The Ugly Duckling

IT was so lovely out in the country! It was summer and the corn stood yellow and the oats green. Down in the green meadows the hay had been stacked, and the stork was walking about there and chattering in Egyptian, for he had learnt that language from his mother. Round the fields and meadows there were vast woods, and in the midst of the woods were deep lakes—yes, it was really lovely out in the country! Right in the sunshine there lay an old manor-house surrounded by deep canals. Great dock-leaves grew from the wall down to the water—they were so tall that small children could stand upright under the biggest of them. They gew like a wild and tangled wood. A duck was sitting on

115

her nest there. She was waiting for her little ducklings to hatch out, but she was rather tired of it now because it had lasted so long and she rarely had a visitor—the other ducks much preferred swimming round in the moat to running up and sitting under a dock-leaf to gossip with her.

At last, one after another, the eggs began to crack. 'Peep, peep!' they said—all the eggs had come alive, and the ducklings were poking their heads out.

'Quack, quack! Hurry up!' she said, and so they made as much haste as they could, and looked all round them under the green leaves, and their mother let them look as much as ever they wanted to, for green is good for the eyes.

'My, how big the world is!' said all the young ones, for they undoubtedly had much more room to move about in now than they had had inside their eggs.

'You don't think this is the whole world!' said their mother. 'Why, it stretches a long way on the other side of the garden, right into the parson's field. Though I have never been so far myself—You're all here now, aren't you?' And so she got up. 'No you're not, the biggest egg is still there— However much longer is it going to be? I'm tired of it now, I can tell you!' And so she sat down again.

'Well, how is it going?' said an old duck who had come to pay her a visit.

'This one egg is taking such a long time,' said the mother-duck. 'It just won't hatch. But you must see the others now! They're the loveliest ducklings I've ever seen, all the image of their father—the wretch, he never comes to visit me!'

'Let me look at that egg that won't crack,' said the old duck. 'You can take it from me, that's a turkey's egg. I was once had, too, in the very same way, and the trouble and bother I had with the young ones! They're afraid of water, I tell you! I could not get them in. I quacked and snapped, but it was no use—Let me see the egg! Yes, it's a turkey's egg right enough. Let it be, and teach the other children to swim.'

'No, I'll sit on it a little longer,' said the duck. 'I've been sitting so long now, I can sit until the Deer Park shuts for the winter!'

'Please yourself!' said the old duck, and off she went.

At last the big egg cracked. 'Peep, peep!' said the young one, as he tumbled out. But oh, how big and ugly he was! The duck looked at him. 'It's a terribly big duckling, that one!' she said. 'None of the others looks like that. Surely it's never a turkey-chick? Well, we shall soon find out. Into the water he shall go, even if I have to kick him in myself!'

The next day the weather was really heavenly, and the sun was shining on all the green dock-leaves. The ducklings' mother came out with all her family and went down to the canal. 'Splash!' She jumped into the water. 'Quack, quack!' she said, and, one after another, the ducklings plopped in. The water closed over their heads, but they came up again at once and floated beautifully. Their legs moved of their own accord, and there they all were, out on the water—even the ugly grey one was swimming with them.

'Well, that's no turkey!' she said. 'Look how beautifully he uses his legs and how straight he holds himself! He's my own child and no mistake. He's really quite handsome if you look at him properly. Quack, quack!—Come with me now, and I'll take you into the world and present you in the duck-yard; but mind you stay close by me all the time in case you get trodden on, and keep a sharp look-out for the cat!'

And so they went into the duck-yard. There was a frightful noise in there, for there were two families fighting over the head of an eel, and then, after all, the cat got it.

'That's the way of the world, you see!' said the ducklings' mother, licking her beak, for she would have liked the eel's head, too. 'Use your legs now!' she said. 'Hurry yourselves along, and bow your heads to that old duck over there! She is the best bred of anyone here. She has Spanish blood—that's why she's so solid-looking, and you see, she has a piece of red cloth tied round her leg. That's something very special, the greatest distinction any duck can have. It means a very great deal indeed—so much that no one will ever get rid of her, and everyone—man and beast alike—treats her with respect. Hurry up now!—Don't turn your toes in! A well-brought-up duckling walks with feet well apart, like father and mother. Now then, bow your heads and say, "Quack!"'

And so they did. But the other ducks round about looked at them and said quite loudly, 'Look at that, we shall have to have that lot as well now, as if there weren't enough of us as it is! Lord, what a queer-looking duckling that one is! We won't put up with him!' And one of the ducks immediately flew at him and pecked him on the back of the neck.

'Let him alone!' said his mother. 'He's not harming anyone!'

'No, but he's too big and queer-looking!' said the duck that had pecked him. 'And so he's got to be put in his place!'

'They're pretty children mother duck has there,' said the old duck with the piece of red cloth round her leg. 'Every one of them pretty, except that one, and he's not turned out at all well. It's a pity she can't make him over again!'

'That can't be done, Your Grace!' said the ducklings' mother. 'He's not good-looking, but he has a very nice disposition, and he swims just as beautifully as any of the others. And what's more, I dare say he'll grow quite handsome, and in time get a little smaller. He's lain too long in the egg, and so he's spoilt his figure.' Then she pecked at the down on the back of his neck and smoothed his body. 'Besides, he's a drake,' she said, 'and so it doesn't matter so much. He'll be strong, I'm sure of it—he'll get along all right!'

'The other ducklings are beautiful,' said the old duck. 'Make yourselves at home now—go and see if you can find an eel's head, and then you can bring it to me!'

And so they made themselves at home.

But the poor duckling that was the last to be hatched out of the egg and was so ugly-looking, was pecked and jostled

and mocked by ducks and hens alike. 'He's too big!' they all said, and the turkey-cock, who had been born with spurs on and therefore thought he was an emperor, puffed himself up like a ship in full sail, went right up to him, and gobbled until he was quite red in the face. The poor duckling did not know where to stand or where to go, he was so miserable at being so ugly and being the laughing-stock of the whole duck-yard.

And so the first day passed, and afterwards it grew worse and worse. The poor duckling was chased about by everyone. Even his brothers and sisters were unkind to him, and kept on saying, 'If only the cat would get you, you ugly thing!' And his mother would say, 'I wish you were far away!' The ducks bit him, and the hens pecked him, and the girl who fed the poultry kicked him with her foot.

Then he ran off and flew away over the hedge. The little birds in the bushes rose terrified into the air. 'That's because I'm so ugly,' thought the duckling, shutting his eyes, but he ran on all the same. Then he came out into the great marsh where the wild ducks lived. There he lay the whole night, he was so tired and unhappy.

In the morning the wild ducks flew up and took a look at their new comrade. 'What kind of a fellow are you?' they asked. The duckling turned from one to another and greeted them as well as he could. ·

'You're ugly all right, and no mistake!' said the wild ducks. 'But that's all the same to us, as long as you don't marry into the family.'—Poor thing, getting married had never entered his head: all he wanted was leave to lie in the rushes and drink a little of the marsh water.

There he lay for two whole days, and then there came two wild geese, or, more properly, two wild ganders, for they were cocks. It was not long since they had been hatched, and so they were very pert and lively.

'Listen, comrade,' they said, 'you're so ugly we've taken quite a liking to you! Will you come along with us and be a bird of passage? Close by here, in another marsh, are some lovely sweet wild geese, young ladies every one of them, and they can say, "Quack!" with the best of them. You're so ugly, you're just the right fellow to try your luck with them!'

'Bang! Bang!' At that very moment there was a loud noise overhead, and both the wild ganders fell down dead in the rushes, and the water became red with blood. 'Bang! Bang!' It sounded again, and great flocks of wild geese flew up out of the rushes. Then the cracking noise was heard yet again. There was a great shoot in progress. The sportsmen lay all round the marsh, some even in the branches of the trees that stretched far out over the rushes. Blue smoke hung like clouds in among the dark trees and far out over the water. The dogs came splashing through the mud; rushes and reeds swayed on all sides. The poor duckling was terrified. Just as he was ducking his head to hide it under his wing, a frightful great dog stood right in front of him, his tongue hanging right out of his mouth and his eyes shining wickedly. He thrust his muzzle right down towards the duckling and bared his sharp teeth—and then—splash!—he had gone again without touching him.

'Oh, thank goodness!' sighed the duckling. 'I'm so ugly that even the dog will think twice before it bites me!'

And so he lay quite still while the duck-shoot whistled in the rushes and shot after shot rang out.

It was far into the day before it was quiet again, but the poor young thing dared not move even then. He waited several hours yet before he looked round him, and then he took to his heels and left the marsh as far behind as he could: he ran over field and meadow, but the wind was blowing so sharply that he found it hard work to get along.

Towards evening he reached a poor little cottage; it was in such a wretched state that it could not make up its mind which side to fall down, and so it remained standing. The wind howled so strongly about the duckling that he had to sit down on his tail to withstand it; and it was growing worse and worse. Then he noticed that the door had fallen off one of its hinges and was hanging so askew that he could creep through the crack into the living-room; and that is what he did.

An old woman lived here with her cat and her hen. The cat, whom she called Sonny, could arch his back and purr; he could make sparks, too, but he had to be stroked the wrong way first. The hen had little short legs, and so she was called Chicky Short-legs. She laid well, and the woman was as fond of her as if she had been her own child.

In the morning the strange duckling was spotted at once, and the cat began to purr and the hen to cluck.

'What's the matter?' said the woman, and looked all round her, but she could not see very well, and so she thought the duckling was a fat duck that had strayed away. 'Here's a fine catch, and no mistake!' she said. 'I can have duck-eggs now, if only it isn't a drake. Well, we shall see!'

And so the duckling was taken on trial for three weeks, but no eggs appeared. Now the cat was the master of the house and the hen was the mistress, and they always said, 'We and the world', for they considered themselves half the world, and the better half at that. The duckling thought there might be other opinions about that, but the hen would not hear of it.

'Can you lay eggs?' she asked.

'No.'

'Then you may hold your tongue!'

And the cat said, 'Can you arch your back and purr and make sparks?'

'No.'

'Then you shouldn't express an opinion when sensible people are talking!'

And the duckling sat in the corner and felt very depressed. Then the thought of fresh air and sunshine came into his mind, and he was seized with such a strange desire to float upon the water that at last he could not help telling the hen about it.

'What's the matter with you?' she asked. 'You've nothing to do, and that's why these odd fancies come over you. Lay an egg or purr, and you'll get over them.'

'But it's so lovely floating on the water,' said the duckling. 'So lovely ducking your head under and diving down to the bottom!'

'Yes, that must be a great pleasure!' said the hen. 'You must have lost your wits! Ask the cat—and he's the cleverest person I know—whether he likes floating on the water or diving to the bottom! I'll leave myself out of it. Ask our mistress, ask the old woman herself—there's no one in the world cleverer than she.—Do you suppose she's seized with a desire to float or duck her head under the water?'

'You don't understand,' said the duckling.

'Well, if we don't understand, then who could! It's certain you will never be cleverer than the cat and the old woman, not to mention myself! Don't show off, child! And thank your Maker for all the good things that have been done for you! Haven't you come into a warm room and found a social background you can learn something from? You talk a lot of nonsense, and your company is not at all amusing. You can believe me! I say it for your own good. I'm telling you a few home-truths about yourself, and that's how you can tell your true friends. Now just see that you lay some eggs and learn to purr or give out sparks!'

'I think I shall go out into the wide world,' said the duckling.

'Well, go on then!' said the hen.

And so the duckling went. He floated on the water and dived below the surface, but all the birds and beasts looked down on him on account of his ugliness.

Now autumn came. The leaves in the woods turned yellow and brown, the wind caught them and whirled them round in a dance. It looked cold up in the sky, and the clouds hung heavy with hail and snow-flakes. The raven, perched on the fence, screeched 'Caw! Caw!' from sheer cold—merely thinking about it was enough to make you feel regularly frozen. The poor duckling certainly did not have a very good time of it.

One evening, as the sun was setting wonderfully, a flock of lovely great birds came out of the bushes. The duckling had never seen anything so beautiful—they were shining white with long graceful necks. They were swans, and uttering a strange noise, they spread their splendid great wings and flew away from those cold parts to warmer lands and open lakes. They rose so high in the air, and the ugly little duckling felt so strange as he watched them; he turned round in the water like a wheel, craning his neck up after them and uttering a cry so loud and strange that it quite frightened him. Oh, he could not forget those lovely birds, those fortunate birds, and as soon as he lost sight of them, he dived right down to the bottom, and when he came up again, he seemed quite beside himself. He did not know what the birds were called nor where they were flying to, and yet he felt more deeply drawn to them than he had ever been to anything. He did not envy them in the slightest—how could it possibly enter his mind to wish himself so beautiful!—he would have been happy if even the ducks had let him stay with them, poor ugly creature that he was!

And the winter grew so very, very cold. The duckling had to swim round and round in the water to keep it from freezing right over, but every night the hole he swam in became smaller and smaller. It froze so hard that the ice-crust creaked and the duckling had to use his legs all the time to prevent the water from icing over. At last, he was tired out and lay quite still, frozen fast in the ice.

Early in the morning a farm-labourer came by, and seeing him, went out and broke the ice up with his wooden clogs, and then carried him home to his wife. There he recovered.

The children wanted to play with him, but the duckling thought they meant to hurt him, and in his fright he flew right into the milk-pan so that the milk splashed out into the room. The woman shrieked, clapping her hands over her head, and then he flew into the tub where the butter was, and then into the barrel of flour and out again.—My, what a sight he was! The woman shrieked and struck at him with the fire-tongs, and the children, laughing and screaming, pushed one another over in their efforts to catch the duckling. It was a good thing the door stood open—he rushed out into the bushes and the new-fallen snow—and there he lay as if in a swoon.

But it would be far too sad a business to describe all the hardship and misery he had to go through during that hard winter. When the sun began to shine warm again, he was lying among the reeds in the marsh. The larks were singing, and it was a lovely spring.

Then one day he stretched his wings. They rustled more strongly than before and bore him swiftly away. Before he was fully aware of it, he was in a great garden where the apple-trees stood in bloom and the lilac hung sweet-scented on its long green boughs right down to the winding canals. Oh, it was beautiful here in the fresh spring! And right in front of him, out of the thick hanging branches, came three lovely white swans, ruffling their feathers and floating lightly on the water. The duckling recognized the magnificent birds and a strange sadness came over him.

'I will fly over to those kingly birds, and they will peck me to death for daring to come near them, I'm so ugly! But it doesn't matter—better to be killed by them than snapped at by the ducks, pecked by the hens, kicked by the girl who looks after the poultry-yard, and suffer misery throughout the winter!' Then he flew out over the water and swam towards the splendid swans. They saw him and came shooting towards him with ruffling feathers. 'Only kill me!' said the poor creature, bowing his head down towards the water and awaiting his death—but what did he see in the clear water? He saw his own reflection beneath him, but he was no longer an awkward, dark-grey bird, ugly and repulsive—he was himself a swan.

It doesn't matter about being born in a duck-yard, as long as you're hatched from a swan's egg.

He felt really glad he had suffered all that hardship and adversity. He could now appreciate his good fortune and all the loveliness that greeted him. And the great swans swam round him and stroked him with their beaks.

Some little children came into the garden and threw bread and corn out into the water, and the youngest cried, 'There's a new one!' And the other children shouted joyfully, 'Yes, there's a new one come!' And, clapping their hands and dancing along, they ran to fetch their father and mother, and bread and cake were thrown into the water, and they all said, 'The new one is the most beautiful of them all. He is so young and lovely!' And the old swans bowed before him.

Then he felt quite shy and hid his head under his wing. He did not know what to do with himself, he was much too happy—but he was not at all proud, for a good heart is never proud. He thought of the time when he had been persecuted and despised, and now he heard them all say he was the loveliest of all those lovely birds. The lilacs bowed their branches right down to the water to him, and the sun was shining warm and fair. Then he ruffled his feathers, lifted up his slender neck, and cried in the joy of his heart, 'I never dreamt I should find so much happiness when I was the ugly duckling!'

The Nightingale

IN China, as of course you know, the Emperor is a
Chinaman, and all those he has about him are
Chinese, too. It happened many years ago now, but
for that very reason the story is worth listening to
before it is forgotten. The Emperor's palace was the
most magnificent in the world, completely and
entirely made of fine porcelain, priceless but so
brittle and fragile that you really had to watch your
step. In the garden the most wonderful flowers
were to be seen, and the most splendid of all had
little silver bells tied to them and they tinkled so
that no one should pass by without noticing them.
Yes, everything was very cunningly contrived in
the Emperor's garden, and it stretched so far that

even the gardener did not know where it ended. If you went on walking, you would come to a lovely forest with lofty trees and deep lakes. The forest went right down to the sea, which was blue and so deep that great ships could sail right in under the branches of the trees. And in them there lived a nightingale that sang so beautifully that even the poor fisherman, who had so many other things to attend to when he was out at night drawing in his nets, would stop work and listen when he heard the nightingale. 'Lord, how beautiful it is!' he would say, but then he would have to attend to his job and the bird would be forgotten. But the next night when it sang again and the fisherman was out, he would say the same thing, 'Lord, how beautiful it really is!'

From all the corners of the world, travellers came to the Emperor's city, and they were filled with admiration for it, and for the palace and garden, too, but when they came to hear the nightingale, they all said, 'That's undoubtedly the best!'

And the travellers talked about it when they got home, and scholars wrote many books about the city, the palace and the garden, but they did not forget the nightingale—that was given first place. And those that could write poetry wrote the loveliest poems all about the nightingale in the forest by the deep sea.

These books went round the world, and one day some of them came to the Emperor, too. He sat in his golden chair and read and read, nodding his head with pleasure every moment at the splendid descriptions of the city, the palace and the garden. Then he came to, 'But the nightingale is undoubtedly the best of all!'

'What's this!' said the Emperor. 'The nightingale! I know nothing of it. Is there such a bird in my empire—and in my own garden of all places? I've never heard of it. What will one read of next!'

So he called for his gentleman-in-waiting, who was so distinguished a person that if anyone of inferior rank dared talk to him or ask him a question, he would answer nothing but, 'P!'—and that, of course, has very little meaning.

'It seems there's a most remarkable bird here called a nightingale,' said the Emperor. 'It's said to be the very finest thing throughout my great empire. Why have I never been told about it?'

'I've never before heard it mentioned,' said the gentleman-in-waiting. 'It's never been presented at court.'

'It's my wish it should come here this evening and sing for me,' said the Emperor. 'It's a fine thing when the whole world knows what I have and I don't!'

'I've never before heard it mentioned,' said the gentleman. 'But I shall search for it. I shall find it.'

But where was it to be found? The gentleman ran up and down all the stairs, through halls and corridors, but not one of all those he met had ever heard the nightingale spoken of. The gentleman ran back again to the Emperor and said it must undoubtedly be a tale invented by those who wrote books. 'Your Imperial Majesty mustn't believe everything that's written—it's most unreliable, and I suspect the Devil has a hand in it!'

'But the book I read it in,' said the Emperor, 'was sent to me from the high and mighty Emperor of Japan, and so it can't be untrue. I will hear the nightingale! And it must be here this evening. It has my very highest favour. And if it doesn't come, the whole court shall be punched in the stomach after they've eaten their supper!'

'Tsing-pe!' said the gentleman and ran once more up and down all the stairs and through all the halls and corridors; and half the court ran with him, for they had no desire to be punched in the stomach. And there they all were, asking after this remarkable nightingale that the whole world knew about but no one at court had ever heard of.

At last they came across a poor little girl in the kitchen. She said, 'Oh, goodness, yes, the nightingale! Of course I know it. My, how it can sing! Every evening they let me take a few of the left-overs from the table home to my poor sick mother. She lives down by the sea-shore, and so when I'm on my way back and feel tired and take a rest in the woods, I hear the nightingale sing. It makes tears come into my eyes, just like mother kissing me.'

'My dear little kitchen-maid,' said the gentleman, 'I'll get you a permanent appointment in the kitchen and leave to see the Emperor eating if only you can take us to the nightingale! Its presence is commanded for this evening.'

So they all made their way out into the forest where the nightingale was in the habit of singing. Half the court was there. And as they were putting their best foot forward, a cow began to low.

'Oh,' said the court page, 'now we've found it! It really has remarkable power for such a small creature. I'm quite certain I've heard it before.'

'No, that's the cows lowing,' said the little kitchen-maid. 'We're still a long way from the place.'

The frogs were now croaking in the pond.

'Lovely!' said the Chinese Master of the Imperial Household. 'Now I can hear her—it's just like tiny church bells!'

'No, that's the frogs,' said the little kitchen-maid. 'But I think we shall soon hear it now.'

And then the nightingale began to sing.

'That's it!' said the little girl. 'Listen! Listen! Look, it's sitting up there!' And she pointed to a little grey bird up in the branches.

'Is it possible!' said the gentleman. 'I never imagined it like that. How ordinary it looks! It's undoubtedly lost its colour at seeing so many distinguished people present.'

'Little nightingale,' cried the little kitchen-maid quite loudly, 'our gracious Emperor would very much like you to sing for him!'

'With the greatest of pleasure,' said the nightingale, and sang so beautifully it was a joy to hear it.

'It's just like glass bells!' said the gentleman. 'And look at the way its little throat works! It's remarkable we've never heard it before. It'll be a great success at court!'

'Shall I sing once again for the Emperor?' said the nightingale, thinking the Emperor was there.

'My excellent little nightingale,' said the gentleman, 'I have great pleasure in inviting you this evening to a celebration at court, where you will charm His High Imperial Grace with your enchanting song.'

'It sounds best out in the green forest,' said the nightingale, but it went willingly with them when it heard that the Emperor wished it.

At the palace everything was properly polished up. The walls and floors, which were of porcelain, shone with the light of many thousand golden lamps. The loveliest flowers, which really could ring, were lined up along the corridors, and running about and the draughts made all the bells ring so loudly you could not hear yourself speak.

Right in the middle of the great hall where the Emperor sat, a golden perch had been placed for the nightingale to sit upon. The whole court was there, and the little kitchen-maid had been given leave to stand behind the door since she now had the title of Royal Kitchen-Maid. All were dressed in their very finest clothes, and they all looked at the little grey bird as the Emperor nodded to it.

Then the nightingale sang so sweetly that tears came into the Emperor's eyes and trickled down over his cheeks, and when the nightingale sang more beautifully still, it went right to his heart and the Emperor was so filled with joy that he said the nightingale should have his golden slipper to wear about its neck. But the nightingale thanked him and said it had already received reward enough.

'I have seen tears in the Emperor's eyes, and that to me is the richest of treasures. An Emperor's tears have a wonderful power. God knows I'm sufficiently rewarded!' And then it sang yet again in its sweet heavenly voice.

'That's the most charming way to flirt I've yet come

across,' said all the ladies in the hall. So they put water in their mouths to make them warble when they were spoken to—they thought they would be nightingales, too. Even the lackeys and the chambermaids let it be known that they were satisfied as well—and that's saying a great deal, for they are the most difficult of all people to please. Yes, indeed, the nightingale was a great success.

It was now to remain at court, where it was to have its own cage and be free to take the air twice during the day and once during the night. Twelve servants were to go with it, each holding fast to a silken cord tied round the bird's leg. There was, of course, no pleasure at all in going out like that.

The whole city talked about the remarkable bird, and whenever two people met, one would say to the other, 'Night!' and the other would reply, 'Gale!' And then they would sigh and understand one another. Yes, and eleven grocers' children were named after it, but not one of them had a note in its body.

One day a large parcel came for the Emperor, and on the outside was written, 'Nightingale'.

'We've a new book here about our famous bird,' said the Emperor. But it was not a book: it was a little mechanical toy lying in a box, a clockwork nightingale made to look like the real one but studded all over with diamonds and rubies and sapphires. As soon as the clockwork bird was wound up, it would sing one of the pieces the real one sang, and its tail, glittering with silver and gold, would go up and down. Round its neck was hung a little ribbon on which was written, 'The Emperor of Japan's Nightingale is a poor thing compared with the Emperor of China's'.

'It's lovely!' they all cried, and the messenger who had brought the mechanical bird was at once given the title of Imperial Nightingale-Bringer-in-Chief.

'Now they can sing together—what a duet that'll be!'

And so they had to sing together, but it did not go right, for the real nightingale sang after its own fashion and the clockwork bird in waltz-time. 'It's not at fault,' said the Master of the Music. 'It keeps excellent time and is quite of my own school.' So the clockwork bird had to sing by itself— it was just as great a success as the real one, and it was so very much more pleasing to look at as well: it glittered like bracelets and brooches.

Three-and-thirty times it sang the selfsame piece, and it was not in the least tired. They would willingly have heard it through again, but the Emperor thought the real nightingale should now sing a little, too—but where was it? It had flown out of the open window away to its green forest, and no one had noticed.

'But what on earth can be the matter!' said the Emperor; and all the courtiers expressed their disapproval of the nightingale and decided that it was an extremely ungrateful creature. 'The one we still have is the better bird, however,' they said, and so the clockwork bird had to sing again. That was the thirty-fourth time they heard the same piece, but they did not quite know it yet, for it was very difficult. The Master of the Music praised the bird beyond all measure and went so far as to assure them that it was better than the real nightingale not only in its plumage which sparkled with so many lovely diamonds, but in its internal mechanism, too.

'For you see, my lords—and, above all, Your Majesty!— with the real nightingale you can never tell what's going to come, but with the clockwork bird you can be absolutely certain! You know it will be thus and not otherwise. You can account for it: you can open it up and reveal the human mind at work; you can see how the waltzes are set, how they are made to play, and how one follows from another.'

'That's exactly what I think,' they all said, and the Master of the Music was given leave to show it to the people the next Sunday. 'They must hear it sing, too,' said the Emperor. And hear it they did. They were all as pleased as if they had got merrily drunk on tea—a regular Chinese habit—and then they all said, 'Oh!', pointed their forefingers (you call them 'lickpots') in the air and nodded. But the poor fisherman who had heard the real nightingale said, 'It sounds pretty enough and there's a likeness, too, but something's missing—though I don't know what.'

The real nightingale was banished from the Emperor's dominions.

The clockwork bird had its own place on a silken cushion close by the Emperor's bed. All the gifts of gold and precious stones it had received lay round about it, and it had been raised to the rank of Singer-in-Chief to the Imperial Bedside-Table. It took precedence as Number One on the Left Side, for the Emperor considered the side where the heart lay the nobler one, and even in an Emperor the heart lies on the left. And the Master of the Music wrote a book in twenty-five volumes about the clockwork bird. It was very scholarly and very lengthy, and contained all the most difficult words in the Chinese language. Everybody said they had read it, for otherwise they would have been thought stupid and punched in the stomach.

And so it went on for a whole year. The Emperor, the court and all the rest of the Chinese people knew off by heart every little trill in the clockwork bird's song, and for that very reason they liked it all the better. They could now sing with it, and they did, too. The street urchins sang, 'Ti-ti-tee! Trill-trill-tri-i-ll!' and the Emperor sang it, too. Yes, there was no doubt about it, it was wonderful.

But one evening when the clockwork bird was singing its best and the Emperor was lying in bed listening to it, something went 'ping!' inside the bird, and it was broken. 'Whir-r-r-r!' The wheels all whizzed round and the music stopped.

The Emperor leaped out of bed at once and called for his physician—but what could he do? So they sent for the clock-maker, and after a great deal of talking and poking about, he got the bird to go after a fashion. But he said it would have to be used very sparingly because the pins were very worn and it was impossible to replace them, without upsetting the music. That was a great blow! Only once a year dared they let the bird sing, and even that was a strain on it. But then the Master of the Music made a little speech full of difficult words and said it was just as good as ever, and so, of course, after that it *was* just as good as ever.

Five years had now passed, and a great sorrow fell upon the whole land; they were all deeply attached to their Emperor, and now he was said to be seriously ill and not expected to live. A new Emperor had already been chosen, and people stood outside in the street and asked the gentleman-in-waiting how their Emperor was.

'P!' he said, shaking his head.

The Emperor lay cold and wan in his magnificent great bed. The whole court thought he was dead, and every one of them ran off to greet the new Emperor. The servants of the bedchamber ran out to gossip about it, and the palace maids were having a big coffee party. Cloth was laid down everywhere in all the halls and corridors so that you should not hear anyone walking about, and it was very, very quiet. But the Emperor was not yet dead. He lay stiff and pale in his magnificent bed with its long velvet hangings and its heavy golden tassles. High up, a window stood open, and the moon shone in upon the Emperor and his clockwork bird.

The poor Emperor could hardly draw breath, and he felt just as if something were sitting on his chest. He opened his eyes, and then he saw that it was Death sitting there wearing his golden crown and holding the Emperor's golden sword in one hand and his splendid standard in the other. And all round him from the folds of the great velvet bed-curtains strange faces peeped out, some quite horrible, others beautiful and gentle. They were all the Emperor's good and evil deeds looking at him now that Death was sitting on his heart.

'Do you remember that?' whispered one after the other. 'Do you remember that?' And then they told him so many things that the sweat broke out on his brow.

'I never knew!' said the Emperor. 'Music, music! Bring the great Chinese drum,' he cried, 'so that I shan't hear all they're saying!'

But on and on they went, and Death nodded like a Chinaman at everything that was said.

'Music, music!' screamed the Emperor. 'Oh, beautiful little golden bird, sing, do sing! I've given you gold and precious things. I've hung my golden slipper about your neck with my own hands. Sing—do, do sing!'

But the bird stood silent: there was no one to wind it up, and without that it could not sing. But Death went on gazing at the Emperor with great empty eye-sockets, and everything was so still, so terribly still.

Then, at that moment, close by the window, the loveliest song rang out. It was the little live nightingale sitting on a branch outside. It had heard of its Emperor's distress, and so had come to bring him comfort and hope. And as it sang, the ghostly forms grew paler and paler, the blood flowed more and more strongly through the Emperor's weak body, and even Death himself listened and said, 'Go on, little nightingale! Go on!'

'Yes, if you will give me that fine golden sword. Yes, if you will give me that rich standard. If you will give me the Emperor's crown!'

And Death gave up each of these treasures for a song, and the nightingale still went on singing. It sang of the quiet churchyard where white roses grow, where the elderberry-tree smells sweet and the fresh grass is damp with the tears of those who are left behind. Then Death was filled with longing for his garden, and drifted out through the window like a cold white mist.

'Thank you, thank you!' said the Emperor. 'You heavenly little bird, I know who you are! I drove you from my realm, and yet you've sung those evil visions from my bed and driven Death from my heart. How shall I reward you?'

'You have rewarded me,' said the nightingale. 'I drew

135

tears from your eyes the first time I sang to you, and I shall never forget it. Those are the jewels that gladden a singer's heart.—But now go to sleep and wake up fresh and strong. I'll sing to you.'

And it sang—and the Emperor fell into a sweet sleep, a sleep that was gentle and refreshing.

The sun was shining in upon him through the windows when he woke up strengthened and in health again. None of his servants had yet returned, for they all thought he was dead, but the nightingale was still sitting there and singing.

'You must stay with me always,' said the Emperor. 'You shall sing only when you want to, and I'll break the clockwork bird into a thousand pieces.'

'No, don't do that,' said the nightingale. 'It's done its best for you. Keep it as you did before. I cannot make my home in the palace, but if you'll let me come when I wish, then in the evening I'll sit outside upon the branch by the window and sing for you to bring you happiness and fill your mind with thoughts. I shall sing of the happy and of those that suffer; I shall sing of the good and evil round about that are hidden from you. The little song bird flies far and wide, to the poor fisherman, to the farm-labourer's cottage, to everyone who is far away from you and your court. I love your heart more than I do you crown, and yet the crown has an air of something holy about it—I'll come and I'll sing for you.— But one thing you must promise me—!'

'Everything!' said the Emperor, standing there in his imperial robes which he had put on himself, and holding the sword, heavy with gold, up to his heart.

'One thing I beg of you. Tell no one that you have a little bird that tells you everything—it will be better if you don't!'

And then the nightingale flew away.

The servants came in to look upon their dead Emperor— yes, there they stood, and the Emperor said, 'Good morning!'

The Swineherd

THERE was once a poor prince. He had a kingdom, which was quite small, yet nevertheless quite big enough to marry on—and that's what he'd set his heart on doing.

Now it would of course be somewhat bold of him to dare to say, 'Will you have me?' to the daughter of the Emperor. But dare he did, for his name was famous far and wide, and there were hundreds of princesses who would have said, 'Yes, please!' But did the Emperor's daughter?

Let s listen to the story and see.

On the grave of the prince's father there grew a rose-tree—oh, such a beautiful rose-tree it was!—which bloomed only once in every five years. And even then it bore only one flower, but that was a rose which smelled so sweetly that its scent would make one forget all one's sorrows and troubles. He had, too, a nightingale, which could sing as if all the beautiful melodies in the world were to be found in its little throat. The princess should have them both, decided the prince. And so the rose and the nightingale were placed in large silver caskets and sent to her.

The Emperor permitted them to be brought before him in the great hall where the princess was playing 'going a-visiting' with her ladies-in-waiting, for that was all they ever did. And when she saw the large caskets with the gifts inside them, she clapped her hands with joy.

'If only it's a little pussy-cat!' she said—but there lay the beautiful rose.

'Oh, but how nicely it's made!' said all the ladies-in-waiting.

'It's more than nice,' said the Emperor. 'It's lovely!'

But when the princess felt it, she was ready to weep.

'Oh, papa!' she said. 'It's not made at all—it's real!'

'Well!' said all the ladies-in-waiting. 'It's real!'

'Now before we get angry,' said the Emperor, 'let's first see what's in the other casket.' And there was the nightingale: it sang so beautifully that for the moment they were unable to say anything bad about it.

'*Superbe! Charmant!*' said the ladies-in-waiting, for they all spoke French, one worse than another.

'How that bird reminds me of her late majesty's musical-box!' said an old courtier. 'Yes, indeed, it has just the same tone, just the same delivery.'

'Yes,' said the Emperor, sobbing like a little child.

'I'll never believe that that's real,' said the princess.

'Indeed, it is—it's a real bird!' said those who had brought it.

'Oh,' said the princess. 'Well, then, let it fly away.'

And nothing would induce her to give the prince permission to visit her.

138

But he would not let himself lose heart. He smeared his face black and brown, pushed his cap down on his head, and knocked at the door.

'Good day, Emperor,' he said. 'I couldn't get a job here in the palace, I suppose?'

'Hm, there are so many after them,' said the Emperor. 'But let me see—I do need someone who can look after the pigs, we've certainly enough of them.'

And so the prince was engaged as Imperial Swineherd. He was given a wretched little room down by the pig-sty, and there he had to stay. All day long he sat working, and when evening came he had made a beautiful little saucepan. It had bells all round it, and as soon as the saucepan came to the boil, they rang clear and sweet, and played the old melody:

> Alas, my dearest Augustin,
> All is now lost, lost, lost!

But the most wonderful thing of all was that, if you held your finger in the steam from the saucepan, you could immediately smell every dinner that was being cooked in every fire-place in the town. That was something very different from a rose!

Now the princess was out walking with all her ladies-in-waiting, and when she heard the melody, she stood quite still and listened with a look of pleasure, for she, too, could play 'Alas, my dearest Augustin!' In fact, it was the only thing she knew, and she played it with one finger.

'Why, that's the piece I know!' she said. 'He must be a very gifted swineherd. Listen, go in and ask him how much he wants for that instrument of his.'

And so one of the ladies-in-waiting had to run in and ask. But she put on her thick shoes first.

'What will you take for that saucepan?' said the lady-in-waiting.

139

'I'll take ten kisses from the princess,' said the swineherd.

'Heaven save us!' said the lady-in-waiting.

'Yes,' said the swineherd, 'but I couldn't take less.'

'Well, what does he say?' asked the princess.

'I really can't say it!' said the lady-in-waiting. 'It's quite dreadful!'

'Then you can whisper,' And so she whispered it.

'He's certainly very rude!' said the princess, and immediately walked away. But when she had walked a few steps the little bells rang out again clear and sweet:

> Alas, my dearest Augustin,
> All is now lost, lost, lost!

'Listen,' said the princess, 'ask him if he'll take ten kisses from my ladies-in-waiting.'

'No, thanks!' said the swineherd. 'Ten kisses from the princess, or else I keep the saucepan.'

'What a tiresome fellow he is!' said the princess. 'But you must all stand round me, so that no one can possibly see.'

And the ladies-in-waiting gathered round her and spread out their dresses; and so the swineherd got the ten kisses, and she got the saucepan.

Well, what pleasure it gave them! The whole of that evening and all the next day, the saucepan was kept boiling. There wasn't a fire-place in the whole town, from the court chamberlain's to the shoemaker's, but they knew what was cooking there. The ladies-in-waiting danced and clapped their hands.

'We know who's going to have sweet soup and pancakes! We know who's going to have porridge and chops! Isn't that interesting?'

'Most interesting!' said the Mistress of the Royal Household.

'Yes, but you must keep it a secret, because I'm the Emperor's daughter.'

'Heavens, of course we shall!' they all said together.

The swineherd—that's to say, the prince, but they knew no differently, of course, and thought he was a real swineherd—didn't let the day go by without busying himself with something. And so he made a rattle: when you swung it round you could hear all the waltzes, barn-dances, and polkas that have ever been known since the creation of the world.

'But that's superb!' said the princess as she was walking by. 'I've never heard a more beautiful piece of music! Listen, go in and ask him the price of that instrument—but I will give no more kisses.'

'He wants a hundred kisses from the princess,' said the lady-in-waiting who had been in to ask.

'He must be mad!' said the princess, and walked away. But when she had gone a few yards she stopped and stood still. 'We must encourage art,' she said, 'and I am the Emperor's daughter. Tell him he may have ten kisses like yesterday; he can have the rest from my ladies-in-waiting.'

'Oh, but we shouldn't like that at all!' said the ladies-in-waiting.

'What nonsense!' said the princess. 'If I can kiss him, so can you. And remember, I provide your board and wages!' And so the lady-in-waiting had to go in and ask him again.

'A hundred kisses from the princess,' he said, 'or we each keep our own.'

'Stand round me!!!' she said. And so all the ladies-in-waiting gathered round her, and he began to kiss her.

'What on earth can that crowd be, down there by the pig-sty!' said the Emperor, as he stepped out on the balcony. He rubbed his eyes and put his glasses on. 'It's the ladies-in-waiting—up to some game, I'll be bound. I'd better go down and see.' And so he pulled his slippers up at the back—for they were really shoes which he had trodden down at the heels.

What a haste he was in, bless him!

As soon as he got down in the yard, he crept up to them very quietly, and the ladies-in-waiting had so much to do with counting the kisses—so that it should be fair, and he shouldn't take too many, nor yet too few—that they were quite unaware of the Emperor. He raised himself up on his toes.

'What on earth!' he said, when he saw them kissing each other. And he struck them both over the head with his slipper just as the swineherd was taking his eighty-sixth kiss. 'Out with you!' said the Emperor, for he was really angry, and both the princess and the swineherd were put outside his empire.

There she stood and wept, the swineherd scolded her, and the rain poured down.

'Oh, what a wretched girl I am!' said the princess. 'If only I'd taken that lovely prince! Oh, how unhappy I am!'

And the swineherd went behind a tree, wiped the black and brown off his face, cast aside his dirty old clothes, and stepped forth again dressed as a prince, so handsome that the princess curtsied as she saw him.

'I've grown to despise you!' he said. 'You wouldn't have an honest prince. You couldn't appreciate the rose and the nightingale, but you could kiss the swineherd for a plaything. Now I wish you well of it!'

And so he went into his kingdom, shut the door and shot the bolt home. So there was nothing for her but to stand outside and sing:

'Alas, my dearest Augustin,
All is now lost, lost, lost!'

The Fir-Tree

OUT in the forest stood a pretty little fir-tree. It grew in a good place where the sun could get to it, and there was plenty of air and the companionship of many larger trees round about, both fir and pine. But the little fir-tree's one thought was to grow. It took no notice of the warm sunshine and the fresh air. It paid no attention to the country children who went by chattering when they were out gathering strawberries or raspberries. They would often pass that way with a whole crock full, or with strawberries threaded on straws, and then they would sit down by the little tree and say, 'Oh, what a pretty little baby tree it is!' But that wasn't at all the kind of thing the tree wanted to hear.

The year after, it was taller by a long new shoot, and the year after that by an even longer one, for you can always tell how many years a fir-tree has been growing by the number of spaces it has between its rings of branches.

'Oh, if only I were a big tree like the others!' sighed the little tree. 'Then I could spread my branches far out all round me and see the wide world from my top. The birds would build their nests among my branches, and when the wind blew I could nod gracefully like the others.'

It found no pleasure in the sunshine or the birds or the red clouds that, morning and evening, sailed overhead.

It was now winter and the snow lay round about white and sparkling, and when a hare came bounding along, it would often jump right over the little tree—and oh, how annoying that was! But two winters went by, and when the third came, the tree was so big that the hare had to go round it. Oh, to grow, to grow, to be big and old!—That, the tree thought, was the only pleasure in the world.

In the autumn the woodcutters came and felled some of the biggest trees. This happened every year, and the young fir-tree, well grown by now, shuddered as the magnificent great trees fell creaking and crashing to the ground. When the branches were hewn off, they looked quite naked as they lay there, long and slender and almost unrecognizable. Then they were laid on wagons, and horses dragged them away out of the forest.

Where were they going? What was in store for them?

In the spring when the swallows and storks returned, the tree asked them, 'Don't you know where they were taken to? Haven't you come across them?'

The swallows knew nothing, but the stork looked thoughtful, nodded his head and said, 'Yes, I think so. I met many new ships as I was flying from Egypt and they all had fine new masts—I dare say they were what you mean. They smelt of fir, and no matter how many times I greeted them, they kept their heads high, very high.'

'Oh, if only I were big enough to fly away over the sea, too!—What actually is the sea? What does it look like?'

'That would take far too long to explain,' said the stork, and off he went.

'Be happy in your youth!' said the sunbeams. 'Be happy in your fresh growth, in the young life that's in you!'

And the wind kissed the tree and the dew wept tears over it, but the fir-tree did not understand them.

As Christmas drew near, quite young trees were felled, trees that were often no bigger and no older than the fir-tree of our story which knew neither peace nor quiet and was always wanting to be off. All these young trees—and they were always the most beautiful—kept their branches. They were laid on wagons, and horses dragged them away out of the forest.

'Where are they going?' asked the fir-tree. 'They're no bigger than I am—in fact, one was much smaller. Why do they keep all their branches? Where are they being taken to?'

'We know! We know!' twittered the sparrows. 'We've peeped through the windows down in the town. We know where they're taken. You can't imagine the brilliance and splendour in store for them! We've peeped in at the windows and seen them planted in the middle of warm rooms and decorated with the loveliest things—gilded apples, ginger-bread, toys and hundreds of lighted candles.'

'And then?' asked the fir-tree, all its branches quivering. 'And then? What happens then?'

'Well, we haven't seen any more. But it was quite wonderful!'

'I wonder if I was born to travel such a brilliant road,' cried the fir-tree excitedly. 'It's even better than sailing over the sea. I'm quite sick with longing. If only it were Christmas! I'm tall now and well formed, with my branches stretching out like the others that were carried away last year.—Oh, if only I were on that wagon! If only I were in that warm room in all that splendour and glory! And then?—Yes, then something even better will follow, something even more beautiful—if not, why should they decorate me like that? Something even greater, even nobler, must come.— But what? Oh, how I suffer! How I long to find out! I don't know myself what's the matter with me.'

'Be happy with me!' said the air and the sunlight. 'Be happy in your fresh youth out in the open!'

But the fir-tree wasn't a bit happy. It grew and grew: winter and summer it stood there, green, dark green it stood. Those that saw it said, 'That's a lovely tree!' and towards Christmas it was the first to be felled. The axe struck deep through its heart, and the tree fell with a sigh to the ground. It felt pain and a sudden weakness. It was quite unable to think of happiness, it was so overcome with sorrow at being parted from its home, the spot where it had sprung up. It knew that it would never again see its dear old companions, or the little bushes and flowers around it, or even, perhaps, the birds. Its journey was by no means pleasant.

The tree was in the courtyard, unloaded with the other trees, before it came to itself and heard a man say, 'That's a fine one! That's the one for us!'

Two men-servants in full livery came and carried the fir-tree into a fine big room. Portraits hung round the walls, and by the great tiled stove stood large Chinese vases with lions on their lids. There were rocking-chairs, silk-covered sofas, great tables full of picture-books and covered with toys worth hundreds and hundreds of pounds—at least, that's what the children said. And the fir-tree was set up in a big tub filled with sand, but no one could see it was a tub because it was draped all round with green cloth and stood on a large brightly-coloured carpet. Oh, how the tree trembled! Whatever would happen now? Men-servants and young ladies came and decorated it: they hung little baskets cut out of coloured paper on the branches, and every basket was filled with sweets. Gilded apples and walnuts were hung up and looked as if they were growing there; and over a hundred little candles, red, blue and white, were fastened firmly to the branches. Dolls that looked really life-like—the tree had never seen anything like them before—swayed among the green needles, and right at the very top of the tree a big gold-tinsel star was fixed. It was magnificent, perfectly magnificent.

'This evening,' they all said, 'this evening it will be a blaze of light!'

'Oh,' thought the tree, 'if only it were evening! If only the candles were lit! And what will happen then, I wonder? I wonder if the trees will come from the forest to look at me? Will the sparrows fly up to the window? Shall I take root here and stand decorated, summer and winter?'

Yes, it could think of nothing else. It had a real barkache from so much longing, and a barkache is just as bad for a tree as a headache is for us.

Now the candles were lit. What radiance! What magnificence! The tree was so excited by it that all its branches quivered and one of the candles set fire to the greenery: the tree sweated with fright.

'Heaven preserve us!' screamed the young ladies as they hastily put it out.

And now the tree dared not give even the smallest quiver. What a dreadful moment it had been! It was so frightened of losing any of its finery and it was quite bewildered by all the brilliance—then the double doors opened wide and a crowd of children burst in as if they would overturn the whole tree. The grown-ups followed soberly. The little ones stood silenced—but only for a moment. Then they shouted for joy till the whole room rang with their noise. They danced round the tree as one present after another was pulled off.

'What is it they're doing?' thought the tree. 'What's going to happen?' The candles burnt right down to the branches, and as they burnt down they were put out. Then the children were given leave to strip the tree. My, they rushed upon it with such violence that all its branches creaked! If it had not been firmly fastened to the ceiling by its tip and the gold star on it, it would have been pushed right over.

The children were dancing around with their lovely toys, and no one looked at the tree except the old nurse who came peeping in among the branches, but that was only to see if a fig or an apple were still left.

'A story! A story!' shouted the children, dragging a stout little man over towards the tree. He sat down right under it. 'Now we're in the greenwood,' he said. 'And oddly enough, it may do the tree good to listen, too. But I'm going to tell only one story. Do you want to hear about Puss in Boots or about Humpty-Dumpty who fell down the stairs but for all that came to the throne and won the princess?'

'Puss in Boots!' screamed some; 'Humpty-Dumpty!' screamed others. There was such a shouting and a screeching, and only the fir-tree remained silent. 'Aren't I going to join in?' it thought. 'Aren't I going to do anything?' But of course it had joined in. It had done all it had to do.

And the man told them about Humpty-Dumpty who fell down the stairs but for all that came to the throne and won the princess. The children clapped their hands and cried, 'Tell us another! Tell us another!' They wanted 'Puss in Boots' as well, but they only got 'Humpty-Dumpty'. The fir-tree stood quite still and thoughtful. The birds out in the forest had never told him anything like that. 'Humpty-Dumpty fell down the stairs and for all that won the princess. Yes, yes, that's how the world goes!' thought the fir-tree, who believed it was all true because it was such a nice man who had told the story. 'Yes, who knows? Perhaps I shall fall down the stairs, too, and win a princess!' And it looked forward to being dressed up again the next day with candles and toys and gold and fruit.

'I shan't tremble tomorrow,' it thought. 'I'll enjoy myself properly in all my glory. Tomorrow I shall hear the story of Humpty-Dumpty again, and perhaps the one about Puss in

Boots as well.' And the tree stood there quiet and thoughtful the whole night through.

In the morning the men-servants and the maids came in.

'Now they'll begin to deck me out in my finery again,' thought the tree, but they dragged it out of the room, up the stairs and into the attic, and there they put it away in a dark corner where no daylight came. 'What's the meaning of this?' thought the tree. 'What am I supposed to do up here? What shall I have to listen to in this place?' And it leant up against the wall, and there it stood and thought and thought—and as the days and nights went by it had plenty of time for thinking. No one came up there, and when at last someone did come, it was only to put some big boxes away in the corner; the tree was now completely hidden and, one would imagine, quite forgotten.

'It's winter outside now,' thought the tree. 'The ground is hard and covered with snow. They can't plant me just yet, and so, no doubt, I'm to stay here under cover until the spring. That is thoughtful. Aren't people kind!—If only it weren't so dark here and so dreadfully lonely!—Not even a little hare!—Yes, it was very pleasant out there in the forest when the snow lay on the ground and the hare scampered by—even when he jumped right over me, though I didn't like it at the time. It really is dreadfully lonely up here.'

At that very moment a little mouse said, 'Squeak, squeak!' and nipped into the room. Then another little mouse followed and together they sniffed round the tree and scurried about among its branches.

'It's horribly cold!' said the little mice. 'But apart from that, it's a lovely place to be in. Isn't it, old fir-tree?'

'I'm not in the least old,' said the fir-tree. 'There are many who are much older than I am.'

'Where do you come from?' asked the mice. 'And what things do you know?' They were very inquisitive. 'Tell us about the loveliest place on earth. Have you been there? Have you been in the pantry where cheeses stand on the shelves and hams hang from the ceiling, where you can dance on tallow-candles, and go in thin and come out fat?'

'I don't know that place,' said the tree. 'But I know the forest where the sun shines and the birds sing.' And so it told them all about when it was young. The little mice had never heard anything like it before—they listened very attentively and said, 'My, what a lot you've seen! You have been lucky!'

'I?' said the fir-tree, and thought over what it had told them. 'Yes, I suppose they really were quite pleasant times.' Then it told them about Christmas Eve when it was decorated with cakes and candles.

'Oh,' said the little mice, 'how lucky you've been, old fir-tree!'

'I'm not old at all!' said the tree. 'It was only this winter I came from the forest. I'm at my very best—I'm just beginning to grow properly.'

'You do tell lovely stories!' said the little mice, and the night after they came with four other little mice to hear what the tree had to tell them, and the more it told them the more clearly it remembered it all and thought, 'Yes, they really were pleasant times. But they may come back, they may come back! Humpty-Dumpty fell down the stairs and yet won the princess. Perhaps I may win a princess, too.' And as it spoke, the fir-tree thought about a pretty little birch-tree that grew out in the forest and, in the eyes of the fir-tree, was a really lovely princess.

'Who's Humpty-Dumpty?' asked the little mice. Then the fir-tree told them the whole story—it could remember every single word. The little mice were ready to jump to the top of the tree with pure delight. The night after that many more mice came, and on Sunday two rats as well. But they said the story wasn't amusing, and that distressed the little mice because now they thought less of it, too.

'Is that the only story you know?' asked the rats.

'The only one,' answered the tree. 'I heard it on the

happiest evening of my life, though at the time I didn't know how happy I was.'

'It's an exceedingly dull story. Don't you know one about bacon and tallow-candles? No pantry-tales?'

'No,' said the tree.

'Well, thank you very much,' answered the rats, and in they went.

At last the little mice stayed away, too, and then the tree sighed, 'It was really very cosy when they sat all round me, those nimble little mice, and listened to what I told them. Now that's gone, too.—But I shall remember to enjoy myself when I'm taken out again.'

But when would that be?—It happened one morning when they came rummaging in the attic. The boxes were moved and the tree was pulled out. They threw it down on the floor—pretty hard, too—and then one of the men dragged it away towards the stairs where the daylight shone.

'Now life's going to begin again,' thought the tree. It felt the fresh air and the first rays of the sun—and then it was out in the yard. Everything happened so quickly that the tree completely forgot to look at itself, there was so much to see all round. The yard adjoined a garden where everything was in bloom: roses hung fresh and sweet-scented over the low fence, lime-trees were in flower, and swallows flew about saying, 'Tweet-tweet-tweet, my husband's come!' But they gave no thought to the fir-tree.

'Now I'm going to live!' it cried joyfully, stretching out its branches—but alas! they were all withered and yellow, and it lay in a corner among the weeds and nettles. The gold paper star, still in place at the tip of the tree, glittered in the bright sunshine.

Two or three of the merry children who had danced round the tree at Christmas and been so delighted with it, were playing in the yard. One of the smallest of them came and tore the gold star off.

'Look what's still left on the ugly old Christmas-tree!' he

151

said, as he trampled on the branches and snapped them under his boots.

The tree looked at the flowers, beautiful and fresh in the garden—it looked at itself, and wished it had stayed in its dark corner in the attic. It thought of its own fresh youth in the forest, of the merry Christmas Eve and of the little mice who had listened with such delight to the story of Humpty-Dumpty.

'Gone! Gone!' said the poor tree. 'If only I'd been happy when I could! Gone! Gone!'

And the man-servant came and chopped the tree into small pieces until a whole pile lay there. It blazed beautifully under the big copper; and it sighed so deeply, each sigh like a pistol-shot, that the children who were playing ran in and, sitting down in front of the fire, looked into it and cried, 'Pop! Pop!' But with each crack—really a deep sigh—the tree thought upon a summer's day in the forest or a winter's night when the stars were shining. It thought of Christmas Eve and Humpty-Dumpty, the only fairy-tale it had ever heard and knew how to tell—and then the tree was burnt up.

The boys played in the yard, and the smallest of them wore upon his chest the gold star which the tree had worn on its happiest evening. It was gone now, the tree was gone, too, and its story was over—over and done with, as all stories are, sooner or later.

The Snow Queen

The First Part: *Which deals with
the mirror and its splinters*

WELL, now, let's begin—and when we come to the end of the story we shall know more than we do now. There was once a demon—one of the very worst—the Devil himself. One day he was in a really good humour because he had made a mirror which had the power of making everything good and beautiful reflected in it disappear almost to nothing, while all that was bad and ugly to look at showed up clearly and appeared far worse than it really was. In this mirror the loveliest of land-scapes looked just like boiled spinach, and even the nicest people looked hideous or else they appeared standing on their heads with no bodies. Their faces were so changed that there was no recognizing them. If anyone had a freckle, you could be certain it would seem to spread all over his nose and mouth. It was great fun, said the Devil.

If a good and holy thought passed through a person's mind, it appeared in the mirror as a grin, and then the Devil had to laugh at his own clever invention. All those who attended the School for Demons—for he ran a School for Demons—spread the news that a miracle had happened. Now for the very first time, they said, you could see what the world and mankind really looked like. They ran everywhere with the mirror, until at last there was not one land or one person that had not been distorted in it. And now they wanted to fly up to Heaven itself to make fun of God and His angels. The higher they flew with the mirror, the louder it laughed, so that they could hardly keep hold of it. Higher and higher they flew, nearer to God and the angels, until at last the mirror shook so violently with laughter that it sprang out of their hands and fell down to earth, where it broke into hundreds of million, billion, and even more, pieces. For that very reason it caused much greater misfortune than before. For some of the pieces were hardly as big as grains of sand, and these flew round about the wide world. Wherever they got into people's eyes, there they stayed, and then the people saw everything distortedly, or else they had eyes only for what was bad in things, for every little splinter of glass had kept the same power that the whole mirror had. Some people even got a little bit of the mirror in their hearts, and then it was really dreadful, for their hearts became just like lumps of ice.

Some pieces of the mirror were so big that they were used as window-panes, but it did not pay to look at your friends through them. Other pieces were used in spectacles, and people had a bad time of it when they put their glasses on to see properly and act justly. The Wicked One laughed until he split his sides, he was so highly tickled by it all. Small splinters of glass were still flying about in the air. So now let us hear what happened.

The Second Part: *A Little Boy and a Little Girl*

In the big city where there are so many houses and people that there is not enough room for everyone to have a little garden, and where, therefore, most people must content themselves with flowers in pots, there were two poor children who had a garden that was a little bigger than a flower-pot. They were not brother and sister, but they were just as fond of one another as if they had been. Their parents lived right next to each other in neighbouring attics. Where the roofs of the two houses met and the gutter ran along under the eaves, the two little windows faced one another, one from each house. All you need do was step over the gutter and you could get from one window to the other.

Their parents had a large wooden box outside each window where they grew vegetables, but in each box there grew also a beautiful little rose tree. Then their parents found that if they placed the boxes across the gutter they reached almost from one window to the other, and looked for all the world like two banks of flowers. As the boxes were very high and the children knew they must not clamber up them, they often got leave to climb out to one another and sit on their little stools under the rose-trees where they played wonderful games together.

In winter such pleasures came to an end. The windows were often completely frosted over, but then the children would warm copper coins on the stove, lay them on the frozen panes and so make lovely peep-holes, perfectly round; and from behind them two sweet gentle eyes would peep out, one from each window. It was the little boy and the little girl. He was called Kay and she was called Gerda. In summer they could be with one another in one jump; in winter they had first to climb all the way downstairs and then all the way upstairs—and outside the snow fell fast.

155

'That's the white bees swarming,' said the old grand-mother.

'Do they have a queen-bee, too?' asked the little boy, for he knew that real bees have one.

'That they have,' said grandmother. 'There she flies, where they are swarming thickest! She is the biggest of them all, and she will never settle quietly upon the ground—up she flies again into the dark clouds. Many a winter night she flies through the streets of the town and peeps in at the windows, and then they frost over with wonderful flower-patterns.'

'Yes, I've seen them,' said both the children, and so they knew it was true.

'Could the Snow Queen come in here?' asked the little girl.

'Just let her try!' said the boy. 'I should put her on the warm stove and then she would melt!'

But grandmother smoothed his hair and told them other tales.

In the evening when little Kay was home and half undressed, he climbed up onto the chairs by the window and peeped out through the little hole. A few snowflakes were falling outside, and one of them, the biggest of them all, came to rest on the edge of one of the window-boxes. The snow-flake grew bigger and bigger, until at last it turned into a lady clothed in the finest white gauze made up of millions of star-like snow-flakes. She was very beautiful, but she was of ice, dazzling, gleaming ice, all through, and yet she was alive; her eyes shone like two clear stars, but there was no rest nor quiet in them. She nodded towards the window and beckoned with her hand. The little boy was terrified and jumped down from the chair; and then it was just as if a great bird flew past the window outside.

The next day there was a clear frost—and then the thaw set in—and then came spring. The sun shone, the green shoots peeped out, the swallows built their nests, the windows were thrown up, and the little children sat once more in their tiny garden high up in the gutter above all the floors of the houses.

The roses bloomed splendidly that summer. The little girl had learnt a hymn, and in it there was a line about roses, and whenever she came to it, she thought about her own. She sang the hymn for the little boy, and he sang it with her:

'In the valley grew roses wild,
And there we spoke with the Holy Child.'

And the little ones held each other's hands, kissed the roses, and, looking up towards God's bright sunshine, they spoke to it as though the Christ Child were there Himself. What beautiful summer days they were; how lovely it was to be outside near the fresh rose-trees that seemed as if they would never stop blooming!

Kay and Gerda sat looking at the picture-book with the animals and birds in it, and it was just at that moment—the clock was striking five in the great church tower—that Kay said, 'Oh! I can feel a pricking in my heart! And now I've got something in my eye!'

The little girl put her arms about his neck and looked; he blinked his eyes, but no, there was nothing to see.

'I think it's gone!' he said; but it hadn't. It was neither more nor less than a splinter of glass from the mirror, the demon-mirror you remember hearing about, that evil glass which made everything great and good reflected in it grow small and ugly, and in which all that was wicked and bad stood out clearly and every blemish became immediately noticeable. Poor Kay—he had a splinter right in his heart as well—it would soon be like a lump of ice. It did not hurt any more—but it was still there.

'What are you crying for?' he asked. 'You look so ugly like that. There's nothing at all the matter with me! Pooh!' he cried in the same breath. 'That rose over there is all worm-eaten. And look, that one's quite lopsided. They really are a disgusting lot of roses—they're just as bad as the boxes they grow in!' And then he kicked the box hard and tore off the two roses.

'Kay, what are you doing?' cried the little girl. And when he saw how frightened she was, he pulled off another rose and sprang in at his own window away from sweet-natured little Gerda.

After that, whenever she came with the picture-book, he said it was only fit for babies. And if grandmother told her tales, he was always ready to criticize—yes, and when he got the chance, he would walk behind her, put her glasses on, and talk just like her. It was a perfect imitation and it made people laugh. He could soon mimic the speech and walk of everybody in the street. Everything that was odd or unpleasant about them, Kay knew how to imitate, and so people said, 'He's certainly got a remarkable head on his shoulders, that boy!' But it was the glass he had got in his eye, the glass that had pierced his heart, and that was why he teased even little Gerda who loved him with all her soul.

His games had become quite different now from what they had been before, they were so intelligent. One winter's day when the snow was falling fast, he came with a large burning-glass, held out the corner of his jacket and let the snow-flakes fall on it.

'Now look in the glass, Gerda!' he said, and every snow-flake grew much bigger and looked like a flower or a ten-pointed star—it was lovely to look at them.

'Do you see how beautifully formed they are,' said Kay. 'They're much more interesting than real flowers! And there isn't a single blemish on them. They're quite perfect—or they would be if they didn't melt!'

Soon afterwards Kay came again, with his big gloves on and his toboggan on his back. He shouted right in Gerda's ear, 'I've got leave to take my toboggan to the big square where the others are playing!' And off he went.

Over on the square the boldest of the boys would often tie their toboggans fast to the farmers' wagons and so ride behind them for quite a long way. It was great fun. When their games were in full swing, a great sledge came by: it was painted white all over, and in it sat a figure muffled in a thick white fur coat and a white fur hat. The sledge drove twice round the square, and Kay quickly managed to tie his little toboggan firmly to it. Then he rode behind it. It went faster and faster, right into the next street. The driver turned round and nodded to Kay in a very friendly fashion, just as if they knew one another. Every time Kay was about to loosen his little toboggan, the driver would nod to him again, and then Kay would remain sitting where he was. They drove right out of the city gates. Then the snow began to fall so thick and fast that the little boy could no longer see his hand in front of him as they sped along. Quickly he undid the rope to free himself from the great sledge, but it was no good—his little tobogggan hung on fast behind, and it went with the speed of the wind. Then he shouted as loudly as he could, but no one heard him, and the snow fell and the sledge flew on. From time to time it gave a bound as if it were flying over ditches and fences. He was completely terrified and wanted to say the Lord's Prayer, but all he could remember were his multiplication-tables.

The snow-flakes grew bigger and bigger until they looked like great white hens. All at once they swerved to one side, the great sledge pulled up, and up stood the driver with fur coat and hat made of pure snow. It was a lady, tall and proud and dazzlingly white—she was the Snow Queen.

'We've made good time,' she said. 'But are you cold? Creep under my bear-skin!' And she seated him beside her in the sledge and drew her fur coat round him—he felt as if he were sinking into a snow-drift.

'Are you still cold?' she asked him, and kissed him on the brow. Oh! her kiss was colder than ice. It went straight to his heart, and his heart was already half-way to being a lump of

159

ice. He felt as if he were about to die—but only for a moment, and then everything was all right again—and he no longer noticed the cold all around him.

'My toboggan! Don't forget my toboggan!' It was the first thing he thought about; it was made fast to one of the white hens, and she flew behind them with the toboggan on her back. The Snow Queen kissed Kay yet once more, and then he forgot little Gerda and grandmother and all the rest of them back home.

'And now you get no more kisses!' she said. 'Or else I shall kiss you to death!'

Kay looked at her. She was very beautiful; a wiser, lovelier face he could not imagine. Now she no longer seemed made of ice as she had done when she sat outside the window and beckoned to him—in his eyes she was perfect. He felt no fear at all. He told her he could do mental arithmetic, even with fractions, and work out how many square miles there were in the country and how many people lived there. And she smiled all the time. Then he realized that he knew very little indeed. He looked up into the great vast vault of the sky, and she flew with him, high up on the dark clouds, and the stormy wind whistled and roared as though it were singing old ballads. They flew over forests and lakes, over land and sea, while below them the cold blast shrieked, the wolves howled, the snow sparkled, and over it flew the black screeching crows. But above them the moon shone large and bright, and Kay gazed upon it all through that long, long winter night. During the day he slept at the Snow Queen's feet.

The Third Part: *The Flower-Garden that belonged to the Old Woman who understood Magic*

But what happened to little Gerda when Kay came back no more? Where could he be? No one knew, no one had any news of him. The boys could only say that they had seen him fasten his little sledge to another, a fine big one, that had driven out into the street and through the city gate. No one knew where he was. Many tears flowed, and little Gerda wept long and sorely. Then they said he was dead, drowned in the river that ran close by the city. Oh, what long, dark winter days those were!

Then the spring came and the sun shone more warmly.

'Kay is dead and gone!' said little Gerda.

'I don't believe it,' said the sunshine.

'He's dead and gone!' she said to the swallows.

'I don't believe it,' they answered, and at last little Gerda did not believe it either.

'I will put my new red shoes on,' she said early one morning, 'the ones Kay has never seen, and then I'll go down and ask the river.'

It was quite early; she kissed her old grandmother who was still asleep, put on her red shoes, and went all alone out of the gate and down to the river.

'Is it true that you've taken my little playmate? I'll make you a present of my red shoes if you'll give him back to me.'

The waves, she thought, nodded very strangely. Then she took her red shoes, her most precious possession, and threw them both out into the river, but they fell close by the bank, and the small waves immediately carried them back to her. It was just as if the river, knowing it had not taken little Kay, would not take her most precious possession either. But Gerda now thought she had not thrown the shoes far enough out, and so she clambered into a boat that lay among the rushes, went right to the farther end of it, and threw the shoes again. But the boat was not made fast, and with her movements it slipped its moorings. Seeing what had happened, she hurriedly tried to regain the bank, but before she could reach it, the boat was a couple of feet out in the water and gliding away ever more swiftly.

Then little Gerda was quite terrified and began to cry,

but no one heard her except the sparrows, and they could not carry her back to land. But they flew along by the bank and sang, as if to comfort her, 'Here we are! Here we are!' The boat drifted downstream. Little Gerda sat quite still in her stockinged feet; her little red shoes floated after her, but they could not reach the boat, which was gathering speed.

It was very beautiful on both banks of the river. There were lovely flowers and old trees and hillsides with sheep and cattle—but not a soul to be seen anywhere.

'Perhaps the river will carry me to little Kay,' thought Gerda. With that she felt more cheerful, and standing up, she watched the beautiful green banks for hours on end. At last she came to a large cherry-orchard where there was a little cottage with odd-looking red and blue windows, a thatched roof besides, and two wooden soldiers standing outside and shouldering arms for those that sailed by.

Gerda shouted to them; she thought they were alive, but, of course, they did not answer. She came quite close to them; the river drove the boat right in towards the shore.

Gerda shouted louder still, and then there came out of the cottage an old, old woman leaning on a crooked stick. She had on a large sun-hat painted all over with the loveliest flowers.

'You poor little child!' said the old woman. 'How on earth did you come to be out on that great strong stream, driven far out into the wide world?' And the old woman waded right out into the water, caught hold of the boat firmly with her crooked stick, drew it to land, and lifted little Gerda out.

Gerda was glad to get back to dry land, though she was a little frightened of the strange old woman.

'Now come and tell me who you are and how you came here,' said the old woman.

And Gerda told her everything; and the old woman shook her head and said, 'Hm! Hm!' And when Gerda had said all and asked her if she had not seen little Kay, the woman said that he had not passed that way yet, but he would come, sure enough. She told Gerda she shouldn't be sad, but should taste her cherries and see her flowers—they were more beautiful than any picture-book and every one of them could tell a whole story. Then she took Gerda by the hand, they entered the little cottage, and the old woman locked the door.

The windows were very high up, and the panes were red, blue, and yellow—the daylight shone strangely inside through all those colours. But on the table stood a dish of the loveliest cherries, and Gerda needed no encouragement to eat as many as she wanted. While she was eating them, the old woman combed her hair with a golden comb, and her hair curled and shone lovely and fair about her friendly little face that was so round and like a rose.

'I've been longing for a dear little girl like you,' the old woman said. 'You shall see now how well we two shall get on together.' And all the time she was combing little Gerda's hair, Gerda was forgetting her foster-brother Kay more and more—for the old woman could cast spells. However, she was by no means a wicked witch and practised magic only a little for her own pleasure, and now she wanted very much indeed to keep little Gerda. So she now went out into the garden, stretched out her crooked stick towards all the rose-trees, and, blooming beautifully though they were, they all sank down into the black earth so that you could not even see where they had been. The old woman was afraid that if Gerda saw the roses she would think of her own, and then remember little Kay and run off.

Then she took Gerda out into the flower-garden. Oh, how fragrant and lovely it was! All the flowers you could think of, flowers of every season of the year, stood there in full splendour—no picture-book could have been more gaily coloured or more beautiful. Gerda jumped for joy and played until the sun went down behind the tall cherry-trees. Then she was given a lovely bed with a red silk eiderdown stuffed with blue violets, and she slept and dreamed as sweetly as any queen upon her wedding-day.

The next day she was allowed to go out again and play among the flowers in the warm sunshine—and thus many days went by. Gerda learnt to know every flower, yet, though there were so many of them, one seemed to be missing, but which one she did not know. Then one day she sat looking at the old woman's sun-hat with its painted flowers, one of which was a lovely rose. The old woman had forgotten to take it off her hat when she made the real ones disappear under the ground. But that's how it is when you're absent-minded.

'What!' cried Gerda. 'Aren't there any roses here?' And she ran among the flower-beds, searching and searching. Her hot tears fell just where a rose-tree had sunk into the earth, and as the tears moistened the ground, the tree shot up at once, just as full of bloom as when it had disappeared; and throwing her arms around it, Gerda kissed the roses and thought about the lovely roses at home—and with them, about little Kay.

'Oh, what a lot of time I've wasted!' said the little girl. 'I was going to find Kay!—Don't you know where he is?' she asked the roses. 'Do you think he's dead and gone?'

'No, he's not dead,' said the roses. 'We've been in the earth where all the dead are, but Kay wasn't there.'

'Oh, thank you, thank you,' said little Gerda, and she went over to the other flowers, and looking into their cups, she asked, 'Don't you know where little Kay is?'

The flowers stood in the sun, each dreaming its own tale. Little Gerda heard many, many things from them, but none of them knew anything of Kay.

What was the tiger-lily saying?

'Can you hear the drum?—Boom, boom! There are only two notes—boom, boom! over and over again. Listen to the women's dirge. Listen to the cry of the priests.—In her long red robe, the Hindu woman stands on the pyre while the flames are mounting round her and her dead husband. But the Hindu woman is thinking of the living there among those that encircle her, of him whose eyes burn hotter than the flames, the fire of whose eyes penetrates her heart more deeply than the flames that will soon burn her body to ashes. Can the heart's flame die in the flames of the pyre?'

'That I don't understand at all!' said little Gerda.

'That is my story!' said the tiger-lily.

What does the convolvulus say?

'High over the narrow field-path hangs an ancient castle. Evergreen plants grow thick and close about the old red walls, leaf upon leaf, right up round the balcony where a lovely girl is standing. She leans out over the parapet and looks down to the path. No rose hanging from its stem is fresher than she, no apple-blossom borne by the wind from the tree is more graceful. How her fine silken gown rustles! "Still he does not come!"'

'Is it Kay you mean?' asked little Gerda.

'I'm speaking only of my own story, my dream,' answered the convolvulus.

What does the little snow-drop say?

'A long plank hangs by a rope between the trees—it is a swing. Two sweet little girls sit swinging—their dresses are as white as snow and long green silk ribbons stream from their hats. Their brother, who is bigger than they, stands up on the swing with his arm about the rope to keep himself steady. He holds a little bowl in one hand and a clay pipe in the other, and he is blowing bubbles. The swing is in motion, and the bubbles fly away with lovely ever-changing colours, the last still clinging to the stem of the pipe and swaying in the breeze. The swing is still in motion. A little black dog, light as the bubbles, jumps up on his hind legs, wanting to join them on the swing. It flies out of his reach, the dog tumbles, barking furiously. They laugh at him, and the bubbles burst. A swinging plank, a fleeting picture of foam— that is my song!'

'What you tell of may well be beautiful, but you speak of

it so sadly and you don't mention little Kay at all. What do the hyacinths say?'

'There were three lovely sisters, delicate and dainty; the first was dressed in red, the second in blue, and the third in pure white. Hand in hand, they danced by the still lake in the clear moonlight. They were not fairy maidens, they were children of men. There was a strong sweet fragrance, and the girls vanished in the woods. The fragrance grew stronger. Three coffins appeared in which the lovely girls lay. They glided from the depth of the dense woods away over the lake. Glow-worms flew shining around them like small flickering lights. Do the dancing girls sleep, or are they dead?—The perfume of the flowers tells they are corpses. The evening bell rings out for the dead!'

'You make me quite sad,' said little Gerda. 'Your scent is so strong it makes me think of those dead girls! Ah, is little Kay really dead then? The roses have been down in the earth, and they say he's not.'

'Ding, dong!' rang the bells of the hyacinths. 'We are not ringing for little Kay—we know nothing of him. We are but singing our own song, the only one we know.'

And Gerda went over to the buttercup, gleaming from among its shiny green leaves.

'You're just like a bright little sun,' said Gerda. 'Tell me, if you can, where shall I find my play-fellow?'

The buttercup shone very prettily and looked at Gerda. What song could the buttercup sing? Not one about Kay at any rate.

'In a little courtyard God's good sun was shining warmly on the first day of spring, its rays gliding down the neighbour's white wall. Close by, the first yellow flowers were growing, gleaming gold in the warm sunbeams. Old grandmother was outside in her chair, and her pretty granddaughter, a poor servant-girl, had come home from a short visit. She kissed her grandmother. There was gold, heart's gold, in that blessed kiss. Gold on lips, and a golden flower; gold

above in the morning hour! There you are, that's my story,'
said the buttercup.

'My poor old granny!' sighed Gerda. 'I know she must be
longing for me—she'll be just as unhappy about me as she
was about little Kay. But I shall soon be home again, and
then I shall bring Kay with me. It's no use asking the flowers.
They know only their own songs—they can tell me nothing!'
And so she tucked up her little frock so that she could run
more quickly. But as she was jumping over it, the narcissus
caught her leg. She stopped short, and looking at the tall
flower, she asked, 'Perhaps you know something?' She bent
right down to it. And what did it say?

'I can see myself! I can see myself!' cried the narcissus.
'Oh, how beautiful my scent is!—Up in her little attic, a little
ballet-dancer is standing, half-dressed, now on one leg, now
on two, and kicking her leg at the whole world—but she is
only an illusion. She pours water from the kettle on to a
piece of cloth she is washing—it is her corset. Cleanliness is a
good thing! Her white dress hangs on its peg—that has been
washed in the kettle too, and put out to dry on the roof. She
puts it on, and arranges her saffron-yellow kerchief about
her neck so that the white of her dress shows brighter than
ever. One leg in the air—look how well she carries herself on
one stalk! I can see myself! I can see myself!'

'I don't care a bit about that,' said Gerda. 'It's not at all
what I want to hear.' And so she ran to the edge of the
garden.

The door was shut, but she twisted the rusty fastening
until it came away and the door flew open, and then little
Gerda ran barefoot out into the wide world. She looked back
three times, but there was no one following her. At last she
could run no more, and so she sat down on a large stone.
When she looked about her, she saw that summer had gone
and it was now late autumn—time had passed unnoticed in
that lovely garden where there was always sunshine and
flowers from every season of the year.

'Oh dear, what a time I've been!' said little Gerda. 'It's autumn already! I dare not rest any longer!' And she rose to go.

Oh, how tired and aching her little feet were! Everything round her looked cold and raw. The long willow-leaves were quite yellow, the wet mist dripped from them, one leaf fell after another. Only the sloe still bore its fruit, so sour and bitter it draws your mouth tightly together. Oh, how grey and heavy it was in the wide world!

The Fourth Part: *Prince and Princess*

Gerda had to rest once more. Right in front of where she was sitting, a big crow was hopping in the snow; he had sat looking at her and wagging his head for a long time, and now he said, 'Caw! Caw!—Goo' daw! Goo' daw!' He could not say it any better, but he meant well towards the little girl and asked where she was going to, out in the wide world all alone. Gerda understood that word 'alone' only too well and felt how deep a meaning it had, and so she told the crow all the story of her life, and asked if he had seen Kay.

The crow nodded thoughtfully and said, 'It could be—it could be!'

'What?—Do you really think you have?' cried the little girl, kissing the crow so hard she very nearly squeezed him to death.

'Careful! Careful!' said the crow. 'I think it may be little Kay. But he's certainly forgotten you now for the princess.'

'Does he live with a princess?' asked Gerda.

'Yes, listen to me,' said the crow. 'But I find it very difficult to speak your language. If you understand crow-talk[1] I can tell you about it much better.'

'No, I'm sorry, I haven't learnt it,' said Gerda. 'But Granny knows it. If only I'd learnt it, too!'

---

1   The Danish *Kragemaal*, literally 'crow-talk', means 'gibberish', like the English 'double-Dutch'. L.W.K.

'Doesn't matter!' said the crow. 'I'll tell you as well as I can, but I shall make a bad job of it all the same.' And so he told her all he knew.

'In this very kingdom where we are now sitting, there lives a princess who's terribly clever, and on top of that, she's read all the newspapers in the whole world—and forgotten them again, she's so clever. A short while ago, as she was sitting on her throne—and that's not as pleasant as you'd think, or so I'm told—she happened to be humming a song, and what should it be but, "Oh, why should I not marry?" "You know, there's something in that!" she says, and so she decided to marry, but she wanted a husband who knew how to answer when you talked to him, not one who just stood and looked distinguished, for she'd soon grow tired of that. Then she had all her ladies-in-waiting summoned by the drum, and when they heard what was in her mind, they were delighted. "Oh, what a good idea!" they said. "I was thinking the same thing, too, only the other day."—Believe me, it's true, every word I'm telling you,' said the crow. 'I've a tame sweetheart who's free to go all over the palace, and she told me everything.'

Of course, she was a crow, too, for birds of a feather flock together, and a crow's mate is always a crow.

'The newspapers came out right away with a border of hearts and the princess's monogram. Any good-looking young man, the newspapers said, was free to go up to the palace and speak with the princess, and the one who talked so that you could hear he felt at home there, and spoke the best, the princess would take for her husband.— Oh, yes,' said the crow, 'believe me, as sure as I'm sitting here, people just streamed in—there was such a crush and commotion! But nothing came of it, either the first day or the second. They could all talk well enough when they were out in the street, but when they came through the palace gate and saw the guard in their silver uniforms and footmen in gold all the way up the stairs and the great halls all lit up, they were dumbstruck. And when they stood before the throne where the princess was sitting, they could find nothing to say except the last word she had spoken herself, and she wasn't really interested in hearing that again. It was just as if everybody there had got snuff on the stomach and had fallen into a stupor until they came out into the street once more—then they could talk all right again. There was a queue right from the city gate to the palace. I was there myself and saw it,' said the crow. 'The men grew both hungry and thirsty, but they didn't get so much as a glass of lukewarm water up at the palace. Some of the most sensible, of course, had brought sandwiches with them, but they didn't share them with their neighbours, for they thought, "If he looks hungry enough, the princess won't have him!"'

'But Kay, little Kay!' asked Gerda. 'When did he come? Was he one of the crowd?'

'Give me time! Give me time! We're coming to him right now. It was the third day, and there came a little fellow, with neither horse nor carriage, who marched coolly up to the palace. His eyes were shining just like yours and he had lovely long hair, but his clothes were shabby.'

'That was Kay!' cried Gerda joyfully. 'Oh, so I've found him!' And she clapped her hands.

'He had a little knapsack on his back,' said the crow.

'Why, that must have been his toboggan,' said Gerda. 'He had his toboggan when he went away.'

'It might have been,' said the crow. 'I didn't look as closely as all that. But I do know from my tame sweetheart that when he came in through the palace gate and saw the Life Guards in their silver uniforms and the footmen in gold all the way up the stairs, he wasn't in the least dismayed—he nodded and said to them, "It must be very dull standing on the stairs—I'd much rather go inside." The halls were blazing with lights; privy councillors and excellencies were walking about barefoot, carrying golden salvers. It was enough to make anyone feel awestruck. His boots squeaked dreadfully, but he wasn't at all afraid.'

'It's certainly Kay!' said Gerda. 'I know he had new boots—I heard them squeaking in Granny's sitting-room.'

'Yes, they squeaked all right,' said the crow. 'And he went quite unconcerned straight in to the princess who was sitting on a pearl as big as a spinning-wheel; and all the ladies-in-waiting with their maids and their maids' maids, and all the gentlemen of the court with their serving-men, their serving-men's serving-men, and their serving-men's serving-men's boys were standing drawn up in order all round them—and the nearer they stood to the door, the haughtier they looked. The serving-man's serving-man's boy, who always went in slippers, looked so haughtily in the doorway, you hardly dare raise your eyes to him!'

'That must have been terrifying!' said little Gerda. 'But did Kay win the princess then?'

'If I hadn't been a crow, I'd have taken her myself—in spite of being engaged. He must have talked just as well as I do when I speak crow-talk, or so my tame sweetheart tells me. He was confident and charming. He had no intention of coming to woo the princess, he had come only to listen to her wisdom, and that he found delightful—and in return, she found him delightful, too.'

'Yes, of course it was Kay!' said Gerda. 'He was so clever, he could do mental-arithmetic with fractions.—Oh, you will take me to the palace, won't you?'

'That's easily said,' replied the crow, 'but how shall we set about it? I must talk it over with my tame sweetheart—she'll no doubt be able to advise us, for I must tell you, a little girl like you would never get leave to enter in the regular way.'

'Oh, yes, I should!' said Gerda. 'When Kay knows I'm here, he'll come straight out to fetch me.'

'Wait for me by the stile there, ' said the crow, wagging his head as he flew off.

The crow did not come back again until dusk had fallen. 'Caw! Caw!' he said. 'I'm to give you my sweetheart's greetings and best wishes. And here's a little loaf for you which she took from the kitchen—they've bread enough there, and you must be hungry.—It won't be possible for you to go into the palace, not with bare feet. The guards in silver and the footmen in gold won't allow it. But don't cry—you shall get in all the same. My sweetheart knows a little back staircase that leads to the bedchamber, and she knows where she can get the key.'

And they made their way into the garden and along the great avenue, where the leaves were falling one after another, and when the lights went out in the palace, one by one, the crow led little Gerda over to a back-door that was standing half-open.

Oh, how Gerda's heart beat with anxiety and longing! She felt just as if she were about to do something wrong, and yet she only wanted to find out if it were little Kay. Yes, it must be he—she thought so vividly about his clever eyes and his long hair that she could actually see the way he used to smile when they sat together under the roses at home. He was sure to be glad to see her, to hear what a long way she had come for his sake, and to know how sad they had all been at home when he did not come back. How frightened she felt, but how glad she was, too!

They were on the stairs now. A little lamp was burning on a cupboard. In the middle of the floor stood the tame crow, turning her head this way and that, and looking intently at Gerda who curtseyed as grandmother had taught her to.

'My young man has spoken very nicely of you, young lady,' said the tame crow. 'And your biography, as they call it, is very touching.—If you will take the lamp, I'll go in front. We're going to take the shortest way, where we shan't meet anyone.'

'I think there's someone coming right behind us now,' said Gerda. Something swished past her, and there seemed to be shadows on the wall—horses with flowing manes and slender legs, huntsmen, lords and ladies on horseback.

'They're only dreams,' said the crow. 'They come and take their lordships' thoughts out hunting—it's a good thing, too, for they can watch them better in bed. But I hope you will show me, if you should come to honour and dignity, that you have a grateful heart.'

'That's hardly something we should talk about,' said the crow from the woods.

They now entered the first hall, where the walls were covered in rose-red satin and beautifully worked flowers. The dreams were already rushing past them, but they swept by so quickly that Gerda failed to see the lords and ladies. Each hall was more magnificent than the last—it was not surprising it took your breath away—and then they were in the bedchamber. Here the ceiling was like a great palm-tree with leaves of glass, costly glass, and in the middle of the floor two beds hung from a thick golden stalk. Each bed was shaped like a lily: one was white, and in it lay the princess; the other was red, and it was there that Gerda must look for little Kay. She turned one of the red petals aside, and she saw the nape of a brown neck— Oh, it was Kay!—She cried his name aloud, holding her lamp towards him—the dreams came rushing on horseback into the room again—he woke, turned his head, and—it was not little Kay.

The prince resembled him only on the back of his neck, but he was young and handsome. And the princess peeped out of the white lily bed and asked what was the matter. Then little Gerda wept and told her whole story and everything the crows had done for her.

'You poor little thing!' said the prince and princess, and they praised the crows and said they were not at all angry with them, but nevertheless they must not do it again. Meanwhile, they should have a reward.

'Would you like to fly away free?' asked the princess. 'Or would you rather have a permanent post as Court-Crows with a right to everything that falls on to the kitchen floor?'

And both the crows bowed low and asked for the permanent post, for they thought of their old age and said, 'It would be a very good thing to have something for the old man,' as they call their declining years.

And the prince got out of his bed and let Gerda sleep in it, and more than that he could hardly do. She folded her little hands together and thought, 'Aren't people kind—and animals, too!' And then she shut her eyes and slept peacefully. The dreams all came flying in again, and this time they looked like God's angels; they pulled a little toboggan behind them, and in it sat Kay nodding to her. But it was all nothing but a dream, and so it vanished as soon as she woke.

The next day she was clothed from top to toe in silk and velvet. She was invited to stay in the palace and enjoy herself, but the only thing she asked for was a little carriage with a horse to pull it and a little pair of boots, and then she could drive out into the wide world again and find Kay.

She was given both the boots and a muff. She was dressed beautifully, and when she was ready to go, a new carriage of pure gold drove up to the door with the prince and princess's coat-of-arms gleaming on it like a star. The coachman, the footmen, and the postilions—there were postilions, too—sat there with golden crowns upon their heads. The prince and princess helped her into the carriage themselves and wished her good luck. The crow from the woods, who was now married, went with her for the first twelve or fifteen miles; he sat beside her, for he could not bear riding backwards. The other crow stood in the gateway flapping her wings; she did not go with them, for, ever since they got their permanent post and too much to eat, she suffered from headaches. The inside of the carriage was lined with sugared cakes, and on the seat were fruits and doughnuts.

'Good-bye! Good-bye!' cried prince and princess, and little Gerda wept and the crow wept.—And so the first few miles went by; then the crow said good-bye, too, and that was the heaviest leave-taking of all. He flew up into a tree and flapped his black wings as long as the carriage, shining like the sun, remained in sight.

The Fifth Part: *The Little Robber-Girl*

They drove through the dark forest, but the carriage shone like a flame, hurting the robbers' eyes until they could bear it no longer.

'It's gold! It's gold!' they shouted, and, rushing out, they seized the horses, struck the little postilions, the coachman, and the footmen all dead, and then dragged little Gerda out of the carriage.

'She's fat, she's a dainty little morsel, she's been fattened on nut-kernels,' said the old hag of a robber-woman with the long bristly beard and the eyebrows that hung down over her eyes. 'She's as good as a little fatted lamb. My, how tasty she'll be!' And she drew out her polished knife which gleamed most terrifyingly.

'Ow!' said the old hag at that very same moment. She had been bitten on the ear by her own little daughter who was hanging on her back and was so wild and out-of-hand it was a joy to see. 'You loathsome brat!' said her mother, and she had no time to kill Gerda.

'She's going to play with me!' said the little robber-girl. 'She's going to give me her muff and her beautiful dress, and sleep in my bed with me.' And then she bit her mother again, so hard that the robber-woman jumped and turned round, and all the robbers laughed and said, 'Look at her dancing with her cub!'

'I'm going to ride in the carriage!' said the little robber-girl, and she insisted upon having her own way, for she was very spoilt and very obstinate. She and Gerda sat inside, and thus they rode over stumps and thorns deeper and deeper into the forest. The little robber-girl was the same height as Gerda, but stronger, broader in the shoulders, and dark-skinned. Her eyes were quite black and had a rather sad look. She put her arm round little Gerda's waist and said, 'They shan't kill you as long as I'm not cross with you. Are you really a princess?'

'No,' said little Gerda and told her everything that had happened to her and how fond she was of little Kay.

The robber-girl gazed at her seriously, nodding her head a little, and said, 'They shan't kill you unless I get really cross with you, and then I shall do it myself!' Then she dried Gerda's eyes, and put both her own hands inside the beautiful muff that was so soft and warm.

And now the carriage stopped. They were in the middle of the courtyard of a robbers' castle. The structure was split from top to bottom, ravens and crows flew out of the gaping holes, and great bull-dogs—every one of them looking as if it could swallow a man—sprang high in the air, but they did not bark, for that was forbidden.

In the great old sooty hall a huge fire was burning in the middle of the flagged floor. The smoke drifted about under the roof and had to find its own way out. A big cauldron of soup was on the boil, and hares and rabbits were turning on the spit.

'You shall sleep here with me tonight, where all my pets are,' said the robber-girl. They got something to eat and drink, and then went over into a corner where straw and rugs lay scattered about. Above their heads, nearly a hundred doves were roosting on laths and perches; they all seemed to be asleep, but they shifted slightly when the little girls came.

'They're all mine, every one of 'em,' said the little robber-girl, promptly seizing one of the nearest, which she held up by the legs and shook so that it beat its wings frantically. 'Kiss it!' she cried, slapping Gerda in the face with it. 'There sit the scum of the forest,' she continued, pointing behind a

176

number of slats nailed across a hole high up in the wall. 'They're forest scum, those two. They'd fly away at once if they weren't properly locked up. And here's my old sweetheart, Bae.' And she dragged a reindeer forward by the horn. It had a bright copper ring round its neck and was tied up. 'We have to keep him under lock and key as well, else he'd run away from us. Every evening of his life I tickle his neck with my sharp knife, and he's ever so frightened of it.' The little girl pulled a long knife out of a crack in the wall and let it glide over the reindeer's neck. The poor beast struck out with its legs, and the robber-girl laughed and pulled Gerda down into the bed.

'Do you have your knife with you when you're going to sleep?' asked Gerda, looking at it somewhat apprehensively.

'I always sleep with my knife!' said the little robber-girl. 'You never know what may happen. But tell me again what you told me before about little Kay, and why you went out into the wide world.' And Gerda told the story from the beginning, and the ring-doves cooed up there in their cage, while the other doves slept. The little robber-girl put her arm round Gerda's neck, and holding her knife in the other hand, she fell asleep, as you could tell by the sound of her breathing. But Gerda was quite unable to shut her eyes, she did not know whether she was to live or die. The robbers sat round the fire, singing and drinking, and the old woman turned head over heels. It was quite terrifying for the little girl to look at.

'Coo, coo!' said the ring doves. 'We've seen little Kay. A white hen was carrying his toboggan, and he was sitting in the Snow Queen's sledge as it swept low over the forest while we lay in our nest. She blew a cold blast on us young ones, and all died except us two. Coo! Coo!'

'What's that you're saying up there?' cried Gerda. 'Which way did the Snow Queen go? Have you any idea?'

'She was doubtless making for Lapland, for there's

always snow and ice there. But you ask the reindeer tethered to that rope.'

'Yes, there's ice and snow, and everything is beautiful and good,' said the reindeer. 'There you can run about freely in the great gleaming valleys. The Snow Queen has her summer tent there, but her stronghold lies up towards the North Pole, on the island called Spitzbergen.'

'Oh, Kay, little Kay!' sighed Gerda.

'You lie still now,' said the robber-girl. 'Or else you'll get my knife in your belly!'

In the morning Gerda told her everything the ring-doves had said, and the little robber-girl looked quite serious, but nodded her head and said, 'Never mind, never mind! —Do you know where Lapland is?' she asked the reindeer.

'Who should know better than I?' answered the animal, his eyes shining at the thought of it. 'That's where I was born and bred and where I used to run over the snow-fields.'

'Listen,' the robber-girl said to Gerda. 'You see all our menfolk are out. The old girl's still here, and here she'll stay, but later on in the morning she'll take a drink from that big bottle and then have a little nap afterwards.—Then I'll see what I can do for you.'

She now sprang out of bed and jumping round her mother's neck, she pulled her beard and said, 'Good morning, my own sweet billy-goat!' Her mother flicked her under the nose and made it red and blue—but it was all done out of pure affection.

So when her mother had had a drink from her bottle and was taking a little nap, the robber-girl went over to the reindeer and said, 'I should love to go on tickling you with my sharp knife because you look so funny when I do it. But never mind, I'm going to undo your rope and help you to get away so that you can run off to Lapland, but you must put your best foot forward and take this little girl to the Snow Queen's palace where her playmate is. I'm sure you overheard what she told me, for she talked loudly enough, and you're always eavesdropping.'

The reindeer jumped for joy. The robber-girl lifted little Gerda up and took the precaution of tying her firmly on and giving her a little cushion to sit on as well. 'Never mind,' she said, 'you can have your fur boots, for it'll be cold; but I'm going to keep the muff—it's much too nice. However, you shan't freeze; you can have mother's big mittens—they'll reach right up to your elbows. Shove your hands in!—now your hands look just like my ugly old mother's.'

And Gerda wept for joy.

'I can't bear to see you snivelling,' said the little robber-girl. 'You must look pleased now. You can have these two loaves and a ham, and then you won't go hungry.' They were tied on to the reindeer's back just behind her. The little robber-girl opened the door, shut all the big dogs up, and then, as she cut through the rope with her knife, she said to the reindeer, 'Off you run now—but see you take good care of the little girl!'

And Gerda stretched out her hands with the mittens on towards the robber-girl and said good-bye, and then the reindeer sped away as fast as he could over bushes and tree-stumps, through the great forest, and over marsh and steppe-land. The wolves were howling and the ravens were screeching. Suddenly there was a crackling in the sky, 'Pop! Pop!' almost as if it were sneezing blood.

'There are my dear old Northern Lights,' said the reindeer. 'Look at the way they're shining!' and with that he ran on even more swiftly, night and day. The loaves were eaten, and so was the ham, and then they found themselves in Lapland.

The Sixth Part: *The Lapp Woman and the Finnish Woman*

They stopped by a little cottage. It was a wretched place with the roof reaching down to the ground and the door so low that the family had to crawl on their stomachs when they wanted to go in or out. There was no one at home but an old Lapp woman who stood grilling a fish over a whale-oil lamp. The reindeer told her the whole of Gerda's story, but he told his own first, for that seemed to be much more important, and Gerda was so overcome with cold she could not speak.

'Ah, you poor, poor things!' said the Lapp woman. 'You've still a long way to go. You'll have to go over four hundred miles into Finmark, for that's where the Snow Queen is staying in the country and burning blue lights every single evening. I'll write a few words on a dried codfish, for paper have I none, and give it to you to take to the Finnish woman up there—she can give you clearer directions than I can.'

And then when Gerda had warmed herself and had something to eat and drink, the Lapp woman wrote a couple of words on a dried codfish, and bidding Gerda take great care of it, she tied her firmly on to the reindeer's back once more, and off he ran. 'Pop! Pop!' came the crackling noise from up in the sky, and all night long the Northern Lights burned a beautiful blue. And then they came to Finmark and knocked on the Finnish woman's chimney-stack, for she had no door at all.

The heat inside was so overpowering that even the Finnish woman went about very nearly naked. She was short and grubby-looking. She at once loosened Gerda's clothing and took off her mittens and boots, otherwise she would have been much too hot. Then she laid a piece of ice on the reindeer's head and read what was written on the codfish. She read it three times, and then when she knew it off by heart, she put the fish into the saucepan for it might just as well be eaten, and she never wasted anything.

Then the reindeer told his own story first and then little Gerda's, and the Finnish woman blinked her eyes wisely but said nothing.

'You're so clever,' said the reindeer, 'I know you can tie up all the winds of the world with a thread of cotton. When the skipper unties the first knot, he gets a good wind; when he unties the second, the wind blows sharp; and when he unties the third and fourth, a gale springs up and the woods come tumbling down. Won't you give the little girl a drink that will give her the strength of twelve men so that she can get the better of the Snow Queen?'

'The strength of twelve men,' said the Finnish woman. 'Yes, that should do the trick.' She went over to a shelf, took a large rolled-up skin from it and unrolled it. Strange letters were written on it, and the Finnish woman read until the sweat poured from her brow.

The reindeer again pleaded so strongly for little Gerda, and Gerda looked at the Finnish woman with such beseeching tearful eyes, that she once again began to blink her eyes and drew the reindeer over into a corner, where she put fresh ice on his head and whispered to him.

'Little Kay is with the Snow Queen right enough, and he finds everything to his liking there and thinks it's the best place on earth, but that's all because he's got a splinter of glass in his heart and a little grain of it in his eye. They must be got out first, otherwise he'll never be human again and the Snow Queen will keep her power over him.'

'But can't you give little Gerda something to take which will give her the power to put everything right?'

'I can't give her greater power than she has already! Can't you see how great that is? Can't you see how she makes man and beast serve her, and how well she's made her way in the world on her own bare feet? She mustn't know of her power from us—it comes from her heart, it comes of her being a sweet innocent child. If she can't find her way into the Snow Queen's palace and free little Kay of the glass splinters all by herself, then we can't help her! Eight or nine miles from here is the beginning of the Snow Queen's garden. You can carry the little girl there. Put her down by the big bush with the

181

red berries which you'll see standing in the snow. Don't stand gossiping, and hurry back!' With that, the Finnish woman lifted little Gerda up on to the reindeer's back, and off he ran as fast as he could.

'Oh, I forgot to bring my boots! And I've left my mittens behind!' cried little Gerda. And she missed them in the stinging cold. But the reindeer dared not stop; he ran on till he came to the big bush with the red berries. There he put Gerda down, and as he kissed her on the lips, great glistening tears ran down the poor animal's cheeks. Then, as fast as he could, he ran back again. There stood poor Gerda, without shoes, without gloves, in the midst of the terrible icy cold of Finmark.

She ran on as well as she could, and then a whole regiment of snow-flakes came towards her. They had not fallen from the sky, for that was quite clear and bright with the Northern Lights. No, the snow-flakes were running along the ground, and the nearer they came, the bigger they grew. Gerda remembered, of course, how large and wonderfully made the snow-flakes had looked when she saw them once through a magnifying-glass. But now they were big and frightening in quite a different way—they were alive, they were the Snow Queen's outposts. They had the strangest shapes. Some looked like nasty great hedgehogs, others like masses of snakes knotted together and darting their heads out, and still others like tubby little bears with bristling hair. All were gleaming white, all were living snow-flakes.

Then little Gerda said the Lord's Prayer, and the cold was so intense she could see her own breath. It rose from her lips like a column of smoke and became thicker and thicker, forming itself into bright little angels that grew bigger and bigger as they touched the ground. They all had helmets on their heads and spears and shields in their hands. There were more and more of them, and when Gerda had finished the Lord's Prayer, there was a whole legion of them round her. They pierced the dreadful snow-flakes with their spears so that they burst into hundreds of pieces, and little Gerda went on her way safe and undismayed. The angels patted her hands and feet, and she felt the cold less keenly and went briskly forward towards the Snow Queen's palace.

But we must first see how Kay is getting on. He was certainly not thinking about little Gerda—least of all that she was standing right outside the palace.

The Seventh Part: *What Happened in the Snow Queen's Palace, and What Happened Afterwards*

The palace walls were of driven snow, the doors and windows of cutting wind. There were over a hundred halls, all as the drifting snow had formed them, the largest stretching for many miles, and all brightly lit by the strong Northern Lights. They were vast, empty, icy-cold, and gleaming. Gaiety never came this way, no, not so much as a little dance for the bears, with the gale blowing up and the polar-bears walking on their hind legs and showing their fine manners; never a little card-party with slap-your-mouth and strike-your-paw; never a little bit of fun over coffee for the young white-fox ladies—empty, vast, and cold it was in the Snow Queen's halls. The Northern Lights flashed with such regularity that you could tell when they had reached their highest point and when they had reached the lowest. In the middle of that empty endless hall of snow there was a frozen lake; it had split into a thousand pieces, and all the pieces were so exactly alike that the whole thing looked like a trick. Whenever she was at home, the Snow Queen sat in the centre of this lake, and then she would say she was sitting in the Mirror of Intelligence, and that it was the best, the only, one in the world.

Little Kay was quite blue with cold—nearly black, in fact—but he did not notice it, for she had kissed his shivers away, and his heart was nothing but a lump of ice. He spent his time dragging sharp flat pieces of ice about, arranging them in all sorts of ways, and trying to make something out of them—it was rather like the kind of thing we sometimes do with small flat pieces of wood when we try to make patterns from them—a Chinese puzzle they call it. Kay made patterns in the same way, most elaborate ones, a sort of intellectual ice-puzzle. In his own eyes the patterns were quite remarkable and of the utmost importance—that was what the grain of glass that was stuck in his eye did for him.

He would lay out his patterns to form written words, but he could never hit upon the way to lay out the one word he wanted, the word 'eternity'. The Snow Queen had said, 'If you can work out that pattern for me, you shall be your own master, and I will present you with the whole world—and a new pair of skates.' But he could not do it.

'Now I must fly off to the warm lands!' said the Snow Queen. 'I must go and peep into the black cauldrons!'—They were the fiery mountains that we call Etna and Vesuvius.—'I must whiten them a little. It's expected of me, and it's good for the lemons and grapes!'

And so the Snow Queen flew away, and Kay sat quite alone in the vast empty hall of ice, many miles in length, and gazed at the pieces of ice, thinking and thinking until his head creaked with the effort. He sat there quite stiff and still: you would have thought he was frozen to death.

It was then that little Gerda stepped into the palace through the great gale of cutting wind; but she said her evening prayers, and the cold winds dropped as if they would fall asleep. She stepped into the vast empty cold halls—then she saw Kay, she recognized him, she flung herself about his neck, held him very tight, and cried, 'Kay! Dear little Kay! So I've found you after all!'

But he sat there quite still, stiff and cold. Then little Gerda wept hot tears that fell upon his breast and penetrated to his heart. They thawed the lump of ice and destroyed the little splinter of glass inside it. He looked at her, and she sang the hymn.

> 'In the valley grew roses wild,
> And there we spoke with the Holy Child!'

Then Kay burst into tears; he wept so desperately that the grain of glass was washed out of his eye. He recognized her and cried joyfully, 'Gerda! Dear little Gerda!—Where have you been all this time? And where have I been?' And he looked round about him. 'How cold it is here! How empty and vast it is!' And as he clung to her she laughed and cried for joy. It was a moment of such bliss that even the pieces of ice danced for joy all round them, and when they grew tired and lay down again, they formed the very letters the Snow Queen had told him he must find out if he were to be his own master and she were to give him the whole world and a new pair of skates.

Gerda kissed his cheeks and the bloom came back to them; she kissed his eyes and they shone like hers; she kissed his hands and feet and he was in perfect health. The Snow Queen could come home when she liked, for his reprieve lay written there in gleaming pieces of ice.

They took one another by the hand, and, as they made their way out of that vast palace, they talked about grandmother and the roses on the roof. And as they walked along, the winds lay still and the sun broke through. When they reached the bush with the red berries, the reindeer was standing there and waiting for them. He had another reindeer with him, a young doe whose udders were full, and she gave the children her warm milk and kissed them on the lips. Then the reindeer carried Kay and Gerda to the Finnish woman's, where they warmed themselves in the heat of the room and were given directions for their journey home, and then to the Lapp woman who had made them new clothes and got her sledge ready.

The reindeer and the young doe ran along by their side and kept them company until they came to the boundary of Lapland. There the first green shoots peeped forth, and there they took their leave of the reindeer and the Lapp woman. 'Good-bye!' they all said. The first little birds began to twitter, there were green buds in the forest, and out of it came riding on a splendid horse—which Gerda recognized, for it had been harnessed to her golden coach—a young girl with a bright red cap on her head and pistols before her. It was the little robber-girl who had grown tired of being at home and had made up her mind to travel northwards first and then, if she did not like it there, in some other direction afterwards. She recognized Gerda at once, and Gerda recognized her. They were very glad to see one another.

'You're a fine sort to go wandering off like that!' she said to Kay. 'I'd like to know whether you deserve to have someone running to the ends of the world for your sake!'

But Gerda stroked her cheek and asked after the prince and princess.

'They've gone travelling to foreign parts,' said the robber-girl.

'And the crow?' asked little Gerda.

'Oh, the crow's dead,' she answered. 'His tame sweetheart's a widow now and goes about with a bit of wool round her leg. She's always moaning and complaining, but it's all put on!—But tell me how you got on, and how you came across him.'

And Gerda and Kay both told her their stories.

'And so that's that!' said the robber-girl. She took them both by the hand, and promised that if she ever passed through their town, she would come up and visit them. Then she rode away out into the wide world, but Kay and Gerda walked on hand in hand.

As they went along, the spring was beautiful with flowers and fresh green leaves. The church bells were ringing, and they recognized the high towers and the great city: it was there that they lived, and on they went till they came to grandmother's door. They went up the stairs and into the living-room, where everything stood in the same place as before, and the clock said 'Tick! Tock!' and the hands turned round. But as they entered the door, they realized that they had grown up. The roses in the gutter were thrusting their flowers in at the open window, and their little stools were still standing there. Kay and Gerda sat down in their own seats and held each other's hands. They had forgotten like a heavy dream the cold empty splendour of the Snow Queen's palace. Grandmother was sitting there in God's bright sunshine and reading aloud from the Bible, 'Except ye become as little children, ye shall not enter into the kingdom of heaven!'

And Kay and Gerda looked into each other's eyes, and all at once they understood the old hymn:

> 'In the valley grew roses wild,
> And there we spoke with the Holy Child!'

There they sat together, grown up, yet children still, children at heart—and it was summer, warm and beautiful summer.

The Red Shoes

THERE was once a little girl who was pretty and dainty, but very poor. In the summer she always had to go barefoot, and in the winter she wore big wooden clogs that made her little ankles quite red—and that looked ugly.

In the middle of the village lived old Grannie Cobbler; she sat and sewed a pair of little shoes as well as she could out of old strips of red cloth—they were very clumsy, but they were well meant and the little girl was to have them. The little girl was called Karen.

She got the red shoes and had them on for the first time the very day her mother was buried. They were hardly the right thing for mourning, but then she had no others, and so with them on her bare feet she walked behind the poor coffin.

Then at that moment a large old carriage came by, and in it there sat a large old lady. She looked at the little girl and felt sorry for her, and so she said to the parson, 'Listen, give that little girl to me and I'll be good to her!'

And Karen thought it was all on account of her red shoes, but the old lady said they were dreadful and so they were burnt. Karen was dressed in neat clean clothes and she had to learn to read and sew. People said she was pretty, but the mirror said, 'You're much more than pretty—you're beautiful!'

Then one day the queen was travelling through the land, and with her she had her little daughter who was a princess. Streams of people came and stood outside the palace, and Karen was there, too. The little princess, dressed in white, stood at a window to let herself be seen. She had neither train nor golden crown, but she did have lovely red morocco-leather shoes—they were quite different from the ones Grannie Cobbler had made for little Karen. There was nothing in the world like red shoes!

Karen was now old enough to be confirmed. She was given new clothes, and she was to have new shoes as well. The wealthy shoemaker in the town took the measurements of her little foot. He did it at home in his own sitting-room where large glass-fronted cupboards stood full of beautiful shoes and shining boots. They looked lovely, but the old lady could not see very well and so she found little pleasure in them. Right in the middle of the shoes stood a pair of red ones, just like those the princess had worn—how beautiful they were! And the shoemaker said they had been made for the daughter of a count but had not fitted.

'They must be made of very highly polished leather,' said the old lady. 'Look at the way they shine!'

'Yes, they shine beautifully!' said Karen. They fitted her and they were bought—but the old lady did not know they were red, for she would never have allowed Karen to be confirmed in red shoes. But that is just what happened.

Everybody looked at her feet, and, as she walked up the aisle towards the choir, even the old pictures on the tombs, portraits of parsons and their wives in starched ruffs and long black gowns, seemed to her to fix their eyes upon her red shoes. They were all she thought about. The parson laid his hand upon her head and spoke to her of holy baptism and her convenant with God. He told her that she must now be a good Christian. The organ played solemnly and the sweet voices of the children were raised in song and the old choir-master sang with them—but the red shoes were all that Karen thought about.

During the afternoon everybody told the old lady that Karen had had red shoes on, and she said it was very naughty and most improper and in future when Karen went to church, she must always go in black shoes, even if they were old ones.

The next Sunday there was communion. Karen looked at the black shoes and she looked at the red ones—and then she looked at the red ones again and put them on. It was lovely bright sunny weather and Karen and the old lady walked along the path through the corn-fields which were rather dusty.

By the church door stood an old soldier with a crutch and a strange long beard; it was more red than white—in fact, it was quite red. He bowed right down to the ground and asked the old lady whether he might dust her shoes. And Karen stretched out her little foot, too. 'What pretty dancing-shoes!' said the soldier. 'Stay fast when you dance!' And as he spoke, he patted the soles of her shoes.

The old lady gave the soldier a small coin and went into the church with Karen.

Everybody present looked at Karen's red shoes, and all the pictures looked at them, and as Karen knelt before the altar and put the golden chalice to her mouth, the red shoes were all she was thinking of and they seemed to be swimming round the chalice in front of her—she forgot to sing the hymn and she forgot to say the Lord's Prayer.

Now they all came out of church, and the old lady stepped into her carriage. As Karen raised her foot to step in after her, the old soldier who was standing close by said, 'What pretty dancing-shoes!' Then Karen could not help dancing a few steps, and once she had begun, her legs went on dancing as if the shoes had some power over them. She danced round the corner of the church and she could not stop. The coachman had to run after her and take hold of her. He lifted her into the carriage, but her feet went on dancing and she kicked the good old lady dreadfully. At last they got her shoes off and her legs lay quiet.

When they got home, the shoes were put away in a cupboard, but Karen could not help going to look at them.

The old lady now lay ill and they said she would not live. She had to be nursed and looked after, and if it was anybody's duty to do this, it was Karen's. But there was a great ball in the town, and Karen was invited. She looked at the old lady, who would not live long anyway: she looked at the red shoes, and thought there could be no harm—she put the red shoes on. She could surely do that!—But then she went to the ball and began to dance.

But when she wanted to turn to the right, the shoes danced to the left; and when she wanted to go up the room, the shoes danced down the room, down the stairs, through the street and out of the town gate. Dance she did and dance she had to, right out into the dark woods.

Then she saw something shining among the trees, and she thought it was the moon for she could see a face, but it was the old soldier with the red beard. He was sitting and nodding his head, and he said, 'Look at those pretty dancing-shoes!'

Then she was frightened and wanted to pull the red shoes

off but they were stuck fast. She tore her stockings off, but the shoes had grown to her feet. And dance she did and dance she had to over field and meadow, in rain and sunshine, by night and by day—but it was most terrifying at night.

She danced into the open churchyard, but the dead there were not dancing—they had something much better to do than dance. She wanted to sit down on the pauper's grave where the bitter tansy grew, but there was neither rest nor peace for her, and, as she danced towards the open church door, she saw an angel there in long white robes and with wings that reached from his shoulders down to the ground. His face was stern and serious and in his hand he held a sword, broad and shining.

'Dance you shall!' he said. 'You shall dance in your red shoes till you grow pale and cold, till your skin shrinks and your body is like a skeleton! You shall dance from door to door and wherever vain, conceited children live, you shall knock that they may hear you and be afraid! Dance you shall, dance—'

'Mercy!' cried Karen. But she did not hear what the angel answered for the shoes carried her through the gate, out into the fields, along highways and byways, and still she had to go on dancing.

Early one morning she danced by a door she knew well. From within came the sound of hymn-singing, and a coffin, decorated with flowers, was carried out. Then she knew the old lady was dead, and she felt herself forsaken by all and cursed by the angel of God.

Dance she did and dance she had to, dance through the dark night. The shoes carried her away through thorn and stubble where she tore herself till the blood ran. Away she danced over the heath to a little lonely house. Here she knew the executioner lived, and she tapped on the window with her fingers and said, 'Come out! Come out!—I can't come in because I'm dancing!'

And the executioner said, 'You surely don't know who I

am? I cut off the heads of wicked men, and I can feel my axe quivering.'

'Don't cut my head off,' said Karen. 'For then I can't repent of my sin. But cut off my feet and the red shoes with them!'

And then she confessed all her sins, and the executioner cut off her feet and her red shoes with them; but the shoes, with her little feet inside them, danced away over the fields into the deep forest.

He carved wooden feet for her and made her a pair of crutches. He taught her a hymn that sinners always sing, and she kissed the hand that had wielded the axe and went away over the heath.

'I've suffered enough now for the red shoes,' she said. 'I'll go to church now so that they can see me!' And she walked as fast as she could towards the church door, but as she came near, the red shoes danced in front of her and she was frightened and turned away.

The whole week through she was sad and shed many heavy tears, but when Sunday came she said, 'There now! Now I've suffered and striven enough. I expect I'm just as good as many of those that sit and hold their heads high in church.' And so she plucked up courage and went, but she got no farther than the gate when she saw the red shoes dancing in front of her. She turned back in terror, repenting her sin deep in her heart.

She went over to the vicarage and asked if she could go into service there. She would work hard and do all she could; she did not mind about wages as long as she had a roof over her head and could live with good people. The parson's wife took pity on her and took her into service, and she was hard-working and grateful. In the evening she would sit quietly and listen to the parson reading aloud from the Bible. The children were all fond of her, but when they talked of fine clothes and dressing up and looking as lovely as a queen, she would shake her head.

The next Sunday, when they were all going to church, they asked her if she would like to go with them, but with tears in her eyes she looked sadly at her crutches, and so the others went to hear God's word while she went alone to her own little room. It was just big enough to hold a bed and a chair, and there she sat down with her hymn-book, and as she was reading in it with a devout heart, the wind carried the sound of the organ over to her from the church. She raised her tear-stained face and said, 'Oh God, help me!'

The sun shone brightly and right in front of her stood the angel of God in his white robes, the one she had seen that night in the church door. But he was no longer holding his sharp sword, instead he carried a lovely green bough full of roses. With it he touched the ceiling and it rose to a great height, and where he had touched it there shone a golden star. He touched the walls and they opened out, and she saw the organ playing, she saw the old pictures of parsons and their wives. The congregation were sitting in the carved and painted pews, singing from their hymn-books.—For the church itself had come home to the poor girl in her narrow little room—or else she had gone to the church. She was sitting in the pew with the rest of the parson's family, and when the hymn came to an end and they looked up, they nodded and said, 'You did right to come, Karen!'

'It was through the grace of God!' she said.

The organ played and the voices of the children in the choir sounded soft and sweet. Bright sunshine streamed warm through the window into the pew where Karen was sitting. Her heart was so full of sunshine, of peace and joy, that it broke. Her soul flew on the sunshine up to God, and there there was no one who questioned her about her red shoes.

The Shepherdess and the Chimney-Sweep

HAVE you ever seen a really old cabinet, black with age and carved into scroll-work and foliage? One just like that used to stand in the living-room, handed down from great-grandmother and carved from top to bottom with roses and tulips. It had the oddest scrolls, and in between them small harts with spreading antlers poked their heads out. But in the middle of the cabinet the figure of a man was carved full length. He was really funny to look at—he was grinning himself: you could hardly call it laughing—for he had billy-goat's legs, small horns on his forehead and a long beard. The children of the house always used to call him 'Field-marshal-major-general-commander-sergeant   Billy-Goat's

Legs' because it was a difficult name to say and there are not many who are given that title—but then it was something of an honour to have him carved there like that. Anyway, there he was, and he was always looking over at the table under the mirror where there stood a beautiful little shepherdess made of porcelain. Her shoes were gilt, her dress prettily looped up with a rose, and besides that she had a golden hat and a shepherd's crook—she was lovely! Close by her stood a little chimney-sweep as black as coal but nevertheless made of porcelain, too. He was just as neat and clean as anyone else, and as for his being a chimney-sweep, that was only something he was supposed to be—the potter could just as well have made a prince of him, for it was all the same to him.

He stood there very charmingly with his ladder, his face pink and white like a girl's—and that was a proper mistake for he ought to have been a little black. He stood quite close to the shepherdess. They had both been put where they stood, and as they were now permanently placed there, they had become engaged. They were well suited to each other— they were both young, made of the same china clay and equally fragile.

Close by them stood yet another figure, three times as big. It was an old Chinaman who could nod his head. He was made of porcelain, too, and said he was the little shepherd-ess's grandfather, but this he could not prove. He insisted, however, that he was responsible for her, and so he had nodded his consent to Field-marshal-major-general-commander-sergeant Billy-Goat's Legs who had proposed to her.

'There you'll have a husband,' said the old Chinaman, 'a husband who, I'm almost certain, is made of mahogany. He can make you Lady-field-marshal-major-general-comman-der-sergeant Billy-Goat's Legs, and he has a whole cupboard full of silver, besides what he has tucked away in secret drawers.'

'I'm not going into that dark cupboard!' said the little shepherdess. 'I've heard tell that he has eleven porcelain wives in there already!'

'Then you can be the twelfth!' said the Chinaman. 'Tonight as soon as the old cupboard creaks you shall be married, as sure as I'm a Chinaman!' And with that he nodded his head and fell asleep.

But the little shepherdess wept and looked at her beloved, the porcelain chimney-sweep.

'I think I'll ask you,' she said, 'to go out into the wide world with me, for we can't stay here.'

'I'll do everything you wish,' said the little chimney-sweep. 'Let's go straight away—I'm sure I can earn you a living by my trade.'

'If only we were safely down from the table!' she said. 'I shan't be happy until we're out in the wide world!'

And he comforted her and showed her where to put her little foot on the carved edges and the gilt foliage round the table leg. He made use of his ladder to help, too, and so at last they were down on the floor, but as they looked over at the old cabinet there was such a commotion—the carved harts were all poking their heads farther out, raising their antlers and turning their necks round as Field-marshal-major-general-commander-sergeant Billy-Goat's Legs sprang high in the air and shouted over to the old Chinaman, 'They're running away! They're running away!'

Then they were a little frightened and quickly jumped into the drawer under the window-seat.

Here lay three or four incomplete packs of cards and a little toy theatre which had been set up as well as possible. A play was going on, and all the queens—diamonds and hearts, clubs and spades—were sitting in the front row fanning themselves with their tulips. The knaves were standing behind them, each with two heads, one at the top and one at the bottom—playing-cards are like that. The play was about two people in love who were prevented from being together, and the shepherdess wept over it because it was just like her own story.

'I can't bear any more of it,' she said. 'I must get out of the drawer!' But when they reached the floor and looked up at the table, the old Chinaman had woken up and was rocking his whole body—the lower part of him was all in one piece.

'The old Chinaman's coming!' screamed the little shepherdess, and she was so distressed she fell right down on her porcelain knees.

'I've an idea,' said the chimney-sweep. 'Let's creep inside that big pot-pourri jar that stands in the corner: we can lie down among the roses and the lavender and throw salt in his eyes if he comes.'

'That won't be any good!' she said. 'Besides, I happen to know the old Chinaman and the pot-pourri jar were once engaged, and when two people have been as close as that there's always some sympathy left between them. No, there's nothing else for it but to go out into the wide world.'

'Have you really enough courage to go out into the wide world with me?' asked the chimney-sweep. 'Have you thought how big it is and that we could never come back here again?'

'Yes, I have!' she said.

The chimney-sweep looked hard at her and then he said, 'The way I'm going is up the chimney. Have you really enough courage to crawl through the stove with me—through the body of the stove and the flue? Then we shall come out into the chimney, and once there I know what I'm about. We shall climb so high they won't be able to reach us, and right at the top there's a hole that leads out into the wide world.'

And he led her across to the stove door.

'It looks very black,' she said, but she went with him, through the body of the stove and the flue, where the darkness was like a pitch-black night.

'Now we're in the chimney,' he said. 'Look! Look! There's the loveliest star shining right over our heads!'

And there really was a star in the sky shining right down on them as if to show them the way. They crawled and they crept—a terrible journey it was, up and up, higher and higher. But he lifted her and helped her up, he held her and showed her the best places to put her little porcelain feet, and so at last they got right up to the top of the chimney, and there they sat down, for they were very tired, as well they might be.

The sky with all its stars was over their heads and below them all the roofs of the town. They could see far and wide all round them, right out into the wide world; the poor shepherdess had never imagined it was anything like that— she laid her little head on her chimney-sweep's shoulder and wept so bitterly that the gilt came off her sash.

'It's too much!' she said. 'I can't bear it! The world is much too big! If only I were back again on the little table under the mirror! I shall never be happy until I'm there again! I've followed you out into the wide world, and now, if you really love me, please take me home again.'

The chimney-sweep spoke to her sensibly, reminding her of the old Chinaman and Field-marshal-major-general-com-mander-sergeant Billy Goat's Legs, but she sobbed so piti-fully and kissed her little chimney-sweep so that he could do nothing but give in to her, foolish though it was.

So they crawled back again with great difficulty down the chimney and they crept through the flue, which wasn't at all pleasant, and at last they found themselves standing in the dark stove. There they hid behind the door to find out how things stood in the living-room. It was quite still. They peeped out—oh dear, there in the middle of the floor lay the old Chinaman. He had fallen down off the table as he attempted to go after them, and now he lay broken into three pieces: the whole of his back had come off in one piece and his head lay where it had rolled into a corner. Field-marshal-

major-general-commander-sergeant Billy-Goat's Legs was standing where he had always stood, and was deep in thought.

'It's dreadful!' said the little shepherdess. 'Old grandfather's broken to bits and it's our fault! I shall never get over it!' And she wrung her tiny little hands.

'He can be riveted, though,' said the chimney-sweep. 'He can be riveted perfectly well.—Now don't get so excited about it! When they've glued his back together and put a good rivet in his neck, he'll be as good as new again and quite able to say all sorts of disagreeable things to us.'

'Do you think so?' she said. And then they climbed up on to the table again where they had stood before.

'Well, we've been a long way,' said the chimney-sweep. 'And we might have spared ourselves the trouble!'

'If only old grandfather were riveted!' said the shepherdess. 'Will it be very expensive?'

But he was riveted. The family had his back glued on, a good rivet was put in his neck, and he was as good as new, but he could no longer nod his head.

'You've grown pretty proud since you were broken to bits,' said Field-marshal-major-general-commander-sergeant Billy-Goat's Legs. 'But it doesn't strike me you've got so much to be proud of. Well, am I going to have her, or aren't I?'

The chimney-sweep and the little shepherdess looked very anxiously at the old Chinaman. They were very much afraid he would nod, but he couldn't, and he found it very disagreeable to tell a stranger that he had a permanent rivet in his neck. And so the porcelain pair were left together, and they blessed grandfather's rivet and loved each other until they broke to pieces.

The High-jumpers

THE flea, the grasshopper and the skipjack once
wanted to see which one of them could jump the
highest, and so they invited the whole world, and
anyone else who wanted to come, to see their show.
Three proper jumpers they looked when they met
together in the living-room.

'Well,' said the King, 'I'll give my daughter to
the one that jumps highest, for it'd be a poor thing
if these persons had to jump for nothing.'

The flea was the first to step forward. He was
very well-mannered and bowed to all sides, for he
had the blood of a young lady in him and was used
to making contact with people, and that, of course,
is a great advantage.

Now came the grasshopper. You could see at once that he was more heavily built, but he had quite a nice way with him and was dressed in a green uniform, the one he was born with. He said, moreover, that he came of a very old family in the land of Egypt and was highly thought of here in Denmark. He had been taken straight from the fields and put into a house of cards, all made of court-cards with the coloured sides turned inwards. It had both doors and windows, and they were cut out of the body of the Queen of Hearts.

'I sing so well,' he said, 'that sixteen crickets, born in the house, that had been chirping away ever since they were small—and still have not been given a house of cards—fretted themselves even thinner than they were before just with listening to me.'

The two of them, both the flea and the grasshopper, gave such good accounts of themselves that they thought they were quite good enough to marry a princess.

The skipjack said nothing, but that meant he was a deep thinker, or so they said. When the Court dog sniffed at him, he guaranteed the skipjack to be of good family. The old councillor, who had been given three Orders for keeping his mouth shut, assured everyone that he knew the skipjack was gifted with second-sight. You could tell by the look of his back whether it was going to be a mild winter or a hard one, and that's more than you can tell from the back of the man that forecasts the weather for the Almanac.

'I'm not saying anything now!' said the King. 'But that's always my way: I keep my thoughts to myself!'

And now it was time for the jumping. The flea jumped so high that no one could see it, and so they insisted that it had not jumped at all, and that was mean of them.

The grasshopper jumped only half as high, but he jumped right into the King's face, and that, the King said, was disgusting.

The skipjack stood hesitating for so long that at last they thought it could not jump at all.

'I hope he's not hurt himself!' said the Court dog, sniffing him again. Then, whoosh!—the skipjack gave a little sideways jump right into the lap of the princess who was sitting low down on a golden stool.

Then the King said, 'The highest jump of all is jumping up to my daughter, and the beauty of it is, you need a good head to think of such a thing. The skipjack has shown us that he has a head—he has legs in his forehead!'

And so he won the princess.

'I jumped the highest all the same,' said the flea. 'But I don't care! Let her have the silly fellow with his stiff legs and nasty ways. I jumped the highest, though, but in this world you need a bit of body to be seen.'

So the flea joined the Foreign Legion, where, they say, he met his death.

The grasshopper sat down outside in a ditch, and thought over the way things really went on in this world, and he said, 'Some body! You need somebody!' Then he sang his own sad song, and that is where we got the story from. But just because it is in print, it doesn't have to be true.

The Little Match-Girl

IT was so dreadfully cold. It was snowing, and the evening was beginning to darken. It was the last evening of the year, too—New Year's Eve. Through the cold and the dark, a poor little girl with bare head and naked feet was wandering along the road. She had, indeed, had a pair of slippers on when she left home; but what was the good of that? They were very big slippers—her mother had worn them last, they were so big—and the little child had lost them hurrying across the road as two carts rattled dangerously past. One slipper could not be found, and a boy ran off with the other—he said he could use it as a cradle when he had children of his own.

So the little girl wandered along with her naked

little feet red and blue with cold. She was carrying a great pile of matches in an old apron and she held one bundle in her hand as she walked. No one had bought a thing from her the whole day; no one had given her a halfpenny; hungry and frozen, she went her way, looking so woe-begone, poor little thing! The snow-flakes fell upon her long fair hair that curled so prettily about the nape of her neck, but she certainly wasn't thinking of how nice she looked. Lights were shining from all the windows, and there was a lovely smell of roast goose all down the street, for it was indeed New Year's Eve— yes, and that's what she was thinking about.

Over in a corner between two houses, where one jutted a little farther out into the street than the other, she sat down and huddled together. She had drawn her little legs up under her, but she felt more frozen than ever, and she dared not go home, for she had sold no matches and hadn't got a single penny, and her father would beat her. Besides, it was cold at home, too. There was only the roof over them, and the wind whistled in, although the biggest cracks had been stopped up with straw and rags.

Her little hands were almost dead with cold. Ah, a little match might do some good. If she only dared pull one out of the bundle, strike it on the wall, and warm her fingers. She drew one out—Whoosh!—How it spluttered! How it burnt! It gave a warm bright flame, just like a little candle, when she held her hand round it. It was a wonderful light. The little girl thought she was sitting in front of a great iron stove with polished brass knobs and fittings; the fire was burning so cheerfully and its warmth was so comforting—oh, what was that! The little girl had just stretched her feet out to warm them, too, when—the fire went out. The stove disappeared— and she was sitting there with the little stump of a burnt-out match in her hand.

Another match was struck. It burnt and flared, and where the light fell upon it, the wall became transparent like gauze; she could see right into the room where the table stood covered with a shining white cloth and set with fine china. There was a roast goose, stuffed with prunes and apples, steaming deliciously—but what was more gorgeous still, the goose jumped off the dish, waddled across the floor with knife and fork in its back, and went straight over to the poor girl. Then the match went out, and there was nothing to see but the thick cold wall.

She struck yet another. And then she was sitting beneath the loveliest Christmas-tree. It was even bigger and more beautifully decorated than the one she had seen this last Christmas through the glass doors of the wealthy grocer's shop. Thousands of candles were burning on its green branches, and gaily coloured pictures, like those that had decorated the shop-windows, were looking down at her. The little girl stretched out both her hands—and then the match went out; the multitude of Christmas-candles rose higher and higher, and now she saw they were the bright stars—one of them fell and made a long streak of fire across the sky.

'Someone's now dying!' said the little girl, for her old granny, who was the only one that had been kind to her, but who was now dead, had said that when a star falls a soul goes up to God.

Once more she struck a match on the wall. It lit up the darkness round about her, and in its radiance stood old granny, so bright and shining, so wonderfully kind.

'Granny!' cried the little girl. 'Oh, take me with you! I know you'll go away when the match goes out—you'll go away just like the warm stove and the lovely roast goose and the wonderful big Christmas-tree.'—And she hastily struck all the rest of the matches in the bundle, for she wanted to keep her granny there, and the matches shone with such brilliance that it was brighter than daylight. Granny had never before been so tall and beautiful, she lifted the little girl up on her arm, and they flew away in splendour and joy, high, high up towards heaven. And there was no more cold and no more hunger and no more fear—they were with God.

But in the corner by the house, in the cold of the early morning, the little girl sat, with red cheeks and a smile upon her lips—dead, frozen to death on the last evening of the old year. The morning of the New Year rose over the little dead body sitting there with her matches, one bundle nearly all burnt out. She wanted to keep herself warm, they said; but no one knew what beautiful things she had seen, nor in what radiance she had gone with her old granny into the joy of the New Year.

The Happy Family

THE largest green leaf in the countryside hereabouts is surely the dock. If you hold one up in front of your little tummy, it makes a proper apron, and if you put one on your head, it is almost as good in rainy weather as an umbrella, for it really is enormously large. The dock never grows by itself. No, where one grows, several others will be growing, too, making a show of great beauty, and all that beauty is food for snails. The big white snails—which, in the old days, gentle folk had made into a fricassee, ate and said, 'Mm, superb!', for they really did think it tasted delicious—well, the snails lived on dock leaves, and so docks were grown especially for them.

Now there was an old manor-house where snails were no longer eaten and so had quite died out. But the docks had not died out. They grew and grew over all the paths and all the beds, and no one could now get the better of them. There was a forest of docks everywhere. Here and there stood an apple-tree or a plum-tree, but apart from that, no one would ever have thought there was a garden there. Everywhere there were docks—and among them lived the two last snails, who were now very old.

They did not know themselves how old they were, but they could well remember that there had been many more of them, that they came of a family from foreign lands, and that the whole forest had been planted especially for them and theirs. They had never been outside it, but they knew there was something else in the world which was called a manor-house. There you were cooked, and then you turned black and were laid on a silver dish, but what happened after that they did not know. Nor could they imagine what it was like to be cooked and to lie on a silver dish, but it must, they thought, be quite delightful and exceedingly grand.

Neither the beetle, nor the toad, nor the earthworm, whom they asked about it, could give them any information. None of them had been cooked or had lain on a silver dish.

The old white snails were the most important beings in the world, a fact they were well aware of. The forest of dock leaves had been brought into existence solely on their account, and the manor-house was there only so that they could be cooked and laid on a silver dish.

They now led a very solitary and happy life, and as they had no children themselves, they had adopted a little common snail that they brought up as their own. The little one would not grow because he was only a common snail, but mother-snail in particular thought she could observe some progress in him. She asked father, since he could not see it, to feel the little one's shell. So of course he felt it and found mother was right.

One day it was raining heavily.

'Listen to it drumming on the dock leaves,' said father-snail.

'And there are drops coming through,' said mother-snail. 'It's running right down the stalk. It'll be wet here, you'll see! I'm glad we've both got a good house, and the little one has his, too. You can't deny that more has been done for us than for all the rest of the creatures. Anyone can see that we are born to be the gentry of the world. We each have a house from the moment of birth, and the forest of docks was sown for our benefit.—I'd like to know how far it stretches, and what there is beyond.'

'There's nothing beyond,' said father-snail. 'Nowhere can be better than where we live, and I wish for nothing more.'

'Well, I do,' said mother. 'I'd like to go to the manor-house and be cooked and laid on a silver dish. All our ancestors were, and there's something very special about it, of that you may be sure.'

'The manor-house has probably fallen down,' said father-snail, 'or else the forest of docks has grown over it and the people can't get out. Anyway, there's no need to be in such a hurry about it, but then you're always in a dreadful rush, and the little one's getting the same way—He's been crawling up that stalk for three days now. I get quite dizzy in the head when I look up at him.'

'Now you're not to be cross with him,' said mother-snail. 'He crawls very carefully. He gives us a great deal of pleasure, and we old folk have nothing else to live for. But have you thought where we are going to get a wife for him? Don't you think somewhere deep in the forest there may be someone of our own sort?'

'There are plenty of black snails, I believe,' said the old chap, 'black snails without houses—slugs, I think they're called. But that would hardly do, and besides, they've fanciful ideas. We could commission the ants to find some-one. They're always running about the place as if they had something to do. They're sure to know of a wife for our little snail!'

'I know the loveliest one you could possibly find!' said the ants. 'But I'm afraid it wouldn't do: she's a queen!'

'That doesn't matter!' said the old snail. 'Has she a house?'

'She has a castle!' said the ant. 'The most beautiful ant-castle with seven hundred corridors.'

'Thanks!' said the mother-snail. 'Our son's not going into an ant-hill!' If you know no better than that, we can give our commission to the white midges. They fly all over everywhere in rain and sunshine; they know the dock forest inside and out.'

'We have a wife for him!' said the midges. 'A hundred man-paces from here, there's a little snail with a house sitting on a gooseberry-bush. She's quite alone and old enough to marry. It's no more than a hundred man-paces.'

'Well, let her come to him', said the old couple. 'He has a forest of docks, and she's only a bush.'

So they fetched little Miss Snail, and it took her eight days to make the journey, and that was the beauty of it, for it showed them that she was of the right sort.

And so the wedding took place. Six glow-worms shone for them as well as they could. Otherwise it all went off very quietly, for the old snails could not stand a lot of drinking and merry-making, but mother-snail made them a very nice speech—father had to leave it to her: he was overcome with emotion—and then they presented them with the whole forest of docks, saying, as they had always done, that it was the best in the world, and if they lived an honest and upright life and had a family, they and their children would one day go to the manor-house, and be cooked and laid on a silver dish.

And when the speech was finished, the old pair crept into their house and never came out again. They went to sleep. The young couple ruled over the forest and had a large family, but they were never cooked and they never lay on a silver dish. And so from that they concluded that the manor-house had fallen down and all the people in the world had died out, and as no one contradicted them, they knew it must be true. The rain beat down on the dock leaves to make drum music for them, the sun shone to give colour to the forest for them, and they were very happy, and the whole family was happy. Indeed, it was.

Everything in its Right Place

IT happened over a hundred years ago now. By a big lake behind the woods stood an old manor-house, and round it were deep ditches where reeds and rushes grew. Close to the bridge that led to the entrance gates an old willow-tree leaned over the reeds.

Over from the sunken road came the sound of horn and hoof-beats. The little goose-girl hastened to drive her geese to one side away from the bridge before the hunting party came galloping along. They rode at such a rate that she had to jump up smartly on to one of the tall stones that stood by the bridge to avoid being ridden down. She was still little more than a child, delicately formed and slightly built, with a happy, contented face and fine clear eyes. But the lord of the manor saw nothing of that. As he came flying past, he tossed up his hunting-crop, caught it the other way on, and with coarse merriment prodded her right in the chest with the handle so that she fell over backwards.

'Everything in its right place!' he cried. 'Into the mire with you!' And then he laughed at his own idea of fun, and the others laughed with him. The whole party was kicking up such a yelling and bawling with the hounds baying in concert that it was indeed a case of: *The rich bird comes tearing along!*—Though God alone knew how rich he really was!

The poor little goose-girl grabbed as she fell and caught hold of one of the hanging branches of the willow. With that she kept herself from slipping into the mud. As soon as the hunt was safely through the gates, she strove to pull herself up, but the branch broke off at the heel, and the goose-girl fell back heavily into the reeds. But at that moment a strong hand grasped her from above. It was a travelling hosier who from a little way off had seen what happened and now came running to help her.

'Everything in its right place!' he said in fun, repeating the squire's words and pulling her up on to the dry ground. He held the broken-off branch up to where it had come from. 'But you can't put everything back in its right place!' he said. And so he pushed it down into the soft earth, saying, 'Grow there if you can, and one day perhaps you'll make a good flute for the folk up at the manor!' He wished the squire and his company a sound thrashing, and went on his way to the manor-house.

However, he did not go to the great hall—he was too humble for that!—but to the servants' hall, where they looked over his wares and bargained with him. From the festive company upstairs came a bawling and roaring that was supposed to be singing. It was the best they could do. There was the sound of laughter and howling dogs. They were over-eating and drinking too much; glasses and jugs were kept brimming with wine and old ale. The hounds ate with them. Now one, now another of the beasts would be kissed by the young gentlemen after its snout had first been wiped on its long ear-flaps. The pedlar was called upstairs with his wares, but only to be made fun of. Wine had gone in and wits had gone out. They poured beer into a stocking for him so that he could drink with them—but he had to do it quickly ! It was all so uncommonly clever and so screamingly funny! Meanwhile, droves of cattle, peasants and farms were being gambled on one card and lost.

'Everything in its right place!' said the hosier when he was once more well away from 'Sodom and Gomorrah', as he called it. 'The open highway, that's my right place. I wasn't in my element up there!' And the little goose-girl nodded to him from the field-gate.

The days and the weeks passed by, and the broken-off willow branch that the hosier had planted down by the ditch was still fresh and green, and had even grown a new shoot. The little goose-girl saw that it must have taken root, and she was delighted, for she thought of it as her own tree.

Yes, that was doing well, but everything else at the manor was going rapidly down-hill with drinking and gambling.

Scarcely six years had passed when the squire, now a poor man with nothing left but his staff and the bag over his shoulder, wandered away from the manor. The house had been bought by a wealthy hosier, the very man who had been made a laughing-stock and offered beer in a stocking. But honesty and enterprise bring prosperity, and the hosier was now lord of the manor, but from that moment there was no card-playing there.

'When the devil first saw the Bible,' he said, 'he decided to have one of his own in mockery of it, and so he invented a pack of cards, and the devil's Bible makes bad reading!'

The new lord of the manor took a wife, and who should

she be but the little goose-girl, who had always been a gentle, pious, good girl. And in her new clothes she looked as fine and beautiful as if she had been born a real lady. How did it happen? Well, that's too long a story for our busy times, but happen it did, and the most important part was yet to come.

Life was pleasant and all went well at the old manor; mother looked after everything inside the house, and father saw to everything outside. Happiness seemed to spring forth there. Prosperity breeds prosperity. The old manor was polished and painted, the ditches were cleaned out and fruit-trees planted. It all looked very comfortable and inviting. The sitting-room floor was as bright and clean as a chopping-board. In the great hall of a winter evening Madame would sit with her maids spinning wool and linen. Every Sunday evening the Bible would be read aloud there by the magistrate himself, for the hosier had been made a magistrate, though not until he was getting on in years. The children grew up there—for they had children—and they were all well educated, though some were not as clever as others, and that, of course, is the case in every family.

The willow branch outside the gates had grown into a fine tree, standing by itself and never cut. 'It's our family-tree!' said the old couple, and they told their children—the stupid ones as well—that they must honour and respect that tree.

And now a hundred years had passed.

We come to our own time. The lake has become a swamp and the old house has been destroyed. A rectangular pond lined with stone is all that remains of the deep ditches, but there still stands near-by a fine old tree with hanging branches—the family-tree—to show how beautiful a willow can be if it is left alone to look after itself. Admittedly the trunk has split down the middle, right from the crown to the roots, and the high winds have twisted it a little, but there it stands, and from all its cracks and crevices where the wind and weather had caused soil to settle, grass and flowers grow. The top especially, where the main branches divide off, looks like a small hanging garden with raspberry canes and chickweed. Even a tiny rowan tree that has rooted there, slender and delicate, in the middle of the old willow, is reflected in the black water whenever the wind drives the duckweed into a corner of the pond. And close by, a little path leads over the home fields.

High up the hill by the woods, where there is a delightful outlook, stands the new manor-house. Large and splendid it is, its window-panes so clear you would think there was no glass in them. The great stone steps that lead up to the front door look like a summer-house with their roses and broad-leaved plants. The lawn is as fresh and green as if every blade of grass were attended to morning and evening. Inside the house, costly paintings hang in the hall; the furniture there is handsome, the chairs and settees are covered in silks and velvets, the tables topped with polished marble, and the books bound in morocco leather tooled in gold-leaf.—No one could doubt that the people who live there are wealthy; and they are distinguished, too—a baronial family.

One thing matched another in its elegance. 'Everything in its right place!' they, too, would say. And so all the pictures which had once held a place of honour in the old house now hung in the passage that led to the servants' quarters. Real rubbish they were, especially two old portraits, the one of a man in a pink coat and a periwig, the other of a lady with powdered hair combed high and a red rose in her hand, but both alike framed in a garland of willow leaves. The two pictures were full of little round holes where the baron's small boys had made a habit of practising their archery on the two old folk, the magistrate and his wife, from whom the whole family had descended.

'But they don't really belong to our family!' said one of the small boys. 'He was a hosier and she was a goose-girl! They were not like papa and mama.'

The pictures were old junk. 'Everything in its right place!' they said, and so great-grandfather and great-grandmother were dismissed to the passage leading to the servants' quarters.

The parson's son was tutor at the manor. He was out walking one day with the small boys and their eldest sister who had recently been confirmed, and they came along the path down towards the old willow-tree. As they walked along, the young lady made a bouquet of field flowers, beautifully arranged with 'everything in its right place'. Nevertheless, she listened very carefully to all that was said, for she loved to hear the parson's son talking of the forces of nature and the great men and women of history. She had a healthy, happy nature, a nobility of thought and spirit, and a heart ready to embrace all of God's creation.

They stopped down by the old willow-tree, and the youngest boy wanted a flute cut, like those he had had before from other willows, and so the parson's son broke off a branch.

'Oh, don't do that!' said the young lady, but it was done. 'That's our famous old tree. I'm very fond of it. They laugh at me for it at home, but I don't mind. There's a tradition about that tree...'

And then she told them all that we have heard about the tree, the old manor, and the goose-girl and the hosier who met there and became the ancestors of that noble family and the young lady herself.

'Those simple, honest old people had no wish to take a title,' she said. 'They had a saying, "Everything in its right place!" and they did not think they would be in their right place if they accepted a barony just because they had money. It was their son, my grandfather, who was made a baron. He was a man of great learning, held a high position, and was well liked by princes and princesses, who invited him to all their functions. He's the one the others at home think most highly of, but for me there's something about the old couple,

218

I'm not sure what, which opens my heart to them. It must have been so comfortable, so patriarchal, in the old house where the mistress sat spinning with her maids, and the old master read aloud from the Bible.'

'They were splendid people, sensible people!' said the parson's son. Then they found themselves discussing the nobility and the middle-classes, the parson's son talking in such a way about being of noble birth as to give the impression of not being himself of middle-class origins.

'Those who belong to a family that has distinguished itself are very fortunate!' said the parson's son. 'It's like having an incentive in one's blood to make the very best of oneself. "Noble" means what it says. It's a gold coin which has not been debased; it has its true value stamped upon it. Among the higher ranks of society you will find many fine acts performed. My mother, for example, was visiting an aristocratic house in the town—my grandmother had, I believe, been nurse to the gracious lady—and was standing in the sitting-room with the noble old lord, when he saw an old woman coming down the street on crutches. She came every Sunday and was given a couple of shillings. "There's that poor old woman," said the lord; "she has such difficulty in getting along!" And before my mother was aware of it, he was out of the door and down the stairs. He, a seventy-year old nobleman, had gone down to the poor woman himself to spare her the difficult climb up the stairs for the few shillings' help she had come for. But when a bit of a fellow, just because he has good blood and a long pedigree, gets up on his hind legs and says in the drawing-room, "There have been people from the gutter here!" after a commoner has been in the room, then that is nobility gone rotten.'

And so the tutor went on at some length, but finally the flute was finished.

There was a big party at the manor with many guests from the district round about and from the capital, ladies dressed with taste and without taste. The great hall was

219

crowded with people. The clergy of the district stood deferentially in a group in a corner giving the impression that they were attending a funeral. It was, however, a celebration that had not yet got going.

There was to be a great concert, and so the baron's little son had brought his willow flute with him into the hall. But he could not blow it, and nor could papa, and so it was no good.

There was music and singing of the sort that usually gives most enjoyment to the performers, but, as it happened, quite pleasant.

'You're a virtuoso, too,' said a young gentleman, who came of a fashionable noble family. 'You play the flute. You cut it yourself. That's true genius: that beats everything.— Sit here on my right—Lord save us! I quite go with the times—one has to, don't you know! You will drive us all to distraction with that little instrument!' And so saying, he handed the parson's son the little flute cut from the willow down by the pond, and he announced loudly and clearly that the tutor would oblige them with a solo on the flute.

The intention was obviously to make fun of him, and so the tutor refused to play, though naturally he could, but they crowded round him and pressed him, and so he took the flute and put it to his mouth.

A strange flute it was! It emitted a note as sustained as the whistle of a steam-engine, but much more powerful. It penetrated through the whole manor, over the gardens, the woods, for miles out into the countryside, and with the sound of it came a great gust of wind roaring, 'Everything in its right place!'—And as if borne by the wind, papa flew out of the house and straight into the cowman's cottage, and the cowman flew—not into the great hall, for that was beyond him—but up into the servants' hall, among all those fine domestics who wore silk stockings, and the haughty fellows were thunderstruck that such a common person should dare sit down to table with them.

Meanwhile in the great hall the baron's young daughter flew up to the top of the table where she was worthy to sit, and the parson's son found himself seated beside her, and there they both sat like a bridal pair. An elderly count from the oldest family in the land remained unshaken in his seat of honour, for the flute was just, and that was as it should be. The witty young gentleman of the fine fashionable parentage who was responsible for the flute-playing flew head-first in among the poultry, and he was not the only one.

The flute was heard for several miles out into the countryside, and sensational events were reported. The family of a wealthy wholesale merchant who were driving four-in-hand were blown completely out of their carriage and could not even find a footing at the back. Two rich farmers who had grown too big for their own corn-fields were blown down into the muddy ditch. It was a dangerous flute! Luckily it split with the first note, and that was a good thing. It was put back into the tutor's pocket: 'Everything in its right place.'

The next day no one spoke of these strange goings-on, and thence came the expression 'to pipe down'! Everything returned to its old place again, except that the two old portraits of the hosier and the goose-girl were found hanging up in the great hall where they had been blown on to the wall. One of the guests who had a real knowledge of art said they had been painted by a master, and so they remained hanging there and were renovated. They had not known before that they were worth anything, and how should they? The paintings now hung in the place of honour—'Everything in its right place.' And that is where everything will be eventually—Eternity is long, longer than this story!

What the Old Man Does is Always Right

I 'M now going to tell you a story that I heard when I was small, and every time I've thought about it since, it's seemed to me much better than before. For it's true of stories, as it is of many people, that they grow nicer and nicer with age—and that's very pleasant!

You've been out in the country, of course, and no doubt you've seen a really old country cottage with a thatched roof overgrown with moss and weeds. There's a stork's nest on the roof-ridge—we can't do without the stork; the walls are uneven and the windows low, and only one of them will open; the baking-oven bulges out like a tubby little tummy, and the elder-bush hangs over the fence

222

where there's a little pond with duck and ducklings right under the gnarled willow-tree. And then, of course, there's the watchdog barking at all and sundry.

There was a cottage just like this out in the country, and two people lived in it, a countryman and his wife. However little they had, they nevertheless had one thing they could do without, and that was a horse that used to go and graze in the ditch alongside the road. Father used to ride to town on it, the neighbours would borrow it and he would get other services in return, but it would still pay them better to sell the horse or exchange it for something or other that would be of more use to them. But what?

'You know best, old man!' said his wife. 'It's market-day in town; ride over and get money for the horse or make a good bargain with it. Whatever you do always turns out right. Now ride off to market.'

And with that she tied his neckerchief, for that at least she could do better than he could. She tied it with a double bow that looked very smart, and then she brushed his hat with the flat of her hand and kissed his warm lips. Then off he rode on the horse that he was going to sell or exchange. Oh, yes, the old man would do that all right.

The sun was burning hot and there wasn't a cloud in the sky. The road was dusty—there were so many people going to market, in carts, on horseback, and on their own feet. It was very hot and there was no shade at all on the road.

A man came along driving a cow, as pretty a cow as you could wish to see. 'She'd give lovely milk for certain sure,' thought the countryman. 'It'd be a real good bargain to get her.

'I say! You with the cow!' he said. 'Shall we have a little chat together? Look, I know a horse costs more than a cow, but I don't mind about that. I could make more use of a cow. Shall we swop?'

'Right you are,' said the man with the cow, and so they swopped.

It was done now, and the countryman could have turned back again, for he had done what he set out to do, but as he had made his mind up to go to market, to market he would go, if only to have a look at it. So on he went with his cow. He walked along briskly, and the cow walked along briskly, and they soon found themselves walking beside a man leading a sheep. It was a good sheep in good condition with a good fleece.

'I'd like to have that,' thought the countryman. 'It'd not want for grazing along the side of our ditch, and in winter we could take it indoors with us. When all's said and done, it'd be more proper for us to keep a sheep than a cow.'

'Shall we swop?' Yes, the man who had the sheep was only too willing, so the bargain was made, and off went the countryman with his sheep along the high road. There by the stile he saw a man with a big goose under his arm.

'That's a fine big 'un you've got there!' said the countryman. 'He's well covered with feathers, and fat, too. He'd look well, tethered by our pond. It'd be something for mother to save her scraps for. She's often said, "If only we had a goose!"—And now she can have one—and she shall and all! Will you swop? I'll give you the sheep for the goose, and a "Thank you" into the bargain!'

The other man was quite willing, and so they swopped and the countryman got his goose. He was near the town now and the road was growing more and more crowded with throngs of men and beasts. They were trudging along the road and overflowing into the ditch right up to the turnpike-keeper's potato-patch where he had tethered his hen so that she shouldn't run away with fright and get lost. She was a short-tailed hen who blinked one eye and looked in good fettle. 'Cluck, cluck!' she said. What she was thinking about, I can't say, but when the countryman saw her, he thought, 'She's the prettiest hen I've ever set eyes on—she's prettier than the parson's brood-hen. I'd like to have her. A hen'll always find a bit of corn. She could almost fend for herself. I think I'd make a good deal of it if I got her for the goose. Shall we swop?' he asked.

'Swop?' said the other. 'Yes, that wouldn't be at all a bad idea!' And so they swopped. The turnpike-keeper got the goose and the countryman got the hen.

He had got through a lot of business on his way to town, it was hot and he was tired. He was in need of a drink and a bite of bread, and there, close at hand, was an inn, but as he was going in, the barman was coming out and ran smack into him in the doorway with a sack crammed full.

'What have you got there?' asked the countryman.

'Rotten apples,' answered the barman. 'A whole sackful of 'em for the pigs.'

'That's a terrible lot. I wish mother could have a sight of that. Last year we'd naught but a single apple on that old tree by the peat-shed. We were going to keep that apple and it stood on the chest of drawers till it burst. "At least it's a crop," says our mother, but she could see a proper crop here. Ah, I wish she might have it!'

'Well, what will you give for it?' asked the barman.

'Give? I'll give you my hen.' And so he gave him the hen in exchange, took the apples and went into the taproom and over to the bar. He leaned his sack of apples up against the stove. The fire was lit, but that hadn't occurred to him. There were many strangers in the room—horse-dealers, cattle-dealers, and two Englishmen who were so rich their pockets were bursting with gold. They were fond of a wager, too, as you shall now hear!

'Hiss-s-s! Hiss-s-s-s!' What was that noise over by the stove? The apples were beginning to roast!

'What's that?' They weren't long finding out. They heard the whole story of the horse that was swopped for the cow, right down to the rotten apples.

'My word, the old woman'll give you a hiding when you get home,' said the Englishmen. 'There'll be the devil to pay!'

'She'll give me no hiding,' said the countryman. 'Our mother'll give me a kiss and say, "What the old man does is always right."'

'Shall we bet on it?' they said. 'Gold coins by the barrel, a hundred pounds to the ton.'

'A bushel will do,' said the countryman. 'I can only lay a bushel of apples, with me and the old woman thrown in— and that's more than flat-measure, that's top-measure!'

'Done!' they said, and the bet was made.

The landlord's cart was brought out. In went the Englishmen, in went the countryman, in went the rotten apples, and off they went to the countryman's cottage.

'Good evening, mother.'

'Thank you, father.'

'Well, I've done a deal.'

'Ah, you're a right good 'un for that,' said his wife, giving him a hug and ignoring the sack and the strangers.

'I swopped the horse for a cow.'

'Thank God for the milk!' said his wife. 'Now we shall have milk to cook with and butter and cheese on the table. That was a real good bargain!'

'Yes, but I swopped the cow for a sheep.'

'But that's even better!' said his wife. 'You're always so thoughtful. We've got just the right grazing for a sheep. Now we can have sheep's milk and sheep's milk cheese and woollen stockings—and woollen nightgowns, too! A cow wouldn't give them. A cow loses her hair. You're a wonderful deep-thinking man.'

'But I swopped the sheep for a goose.'

'What, my old dear, shall we really have a goose for Martinmas this year? You're always thinking up something to please me. That was a lovely thought! We'll tether the goose and it'll be fatter still by Martinmas.'

'But I swopped the goose for a hen,' said her husband.

'A hen! That was a good bargain!' said his wife. 'The hen'll lay eggs and hatch 'em, and then we shall have chickens and a proper poultry-yard. That's just what I've been longing for.'

'Yes, but I swopped the hen for a sack of rotten apples.'

'Oh, now I really shall have to give you a kiss!' said his wife. 'Oh, thank you, thank you, my own dear husband! Now I must tell you something. After you'd gone, I thought I'd make a real nice supper for you, an omelette with chives in it. I had the eggs, but I'd got no chives. So I went over to the schoolmaster's. I know they've got chives there, but the wife's an old skinflint and so I asked her just to lend me some. "Lend you some?" she said. "Nothing grows in our garden, not even a rotten apple!" There I was and I couldn't lend her one in return. But now I can lend her ten—a whole sackful, if need be! That's real funny, that is, father!' And with that she kissed him right on the mouth.

'That's what I like to see!' said the Englishmen. 'Going down hill all the time and just as pleased as ever. It's worth the money.' And so they paid a ton of gold coins to the countryman who had got a kiss instead of a hiding.

Yes, it always pays when the wife realizes and makes it clear to others that the old man is the wise one and what he does is right.

Well, there's the story. I heard it when I was small, and now you've heard it, too, and you know that what the old man does is always right.

The Goblin at the Grocer's

THERE was a student—an unmistakable student—
who lived in the attic and owned nothing. There
was a grocer—equally unmistakable—who lived on
the ground-floor and owned the whole house. The
goblin attached himself to the grocer, for every
Christmas Eve the grocer gave him a dish of
porridge with a great lump of butter in it. The
grocer could easily spare it, and so the goblin
stayed in the shop—and there was much to be
learnt there.

One evening the student came in by the back
door to buy himself some candles and cheese—he
had no one to send and so he came himself. He got
what he wanted and paid for it, and the grocer and

his wife nodded 'good evening'—the wife was a woman who could do more than nod, for she had the gift of the gab!—and the student nodded back and then stood reading the bit of paper the cheese was wrapped in. It was a page torn out of an old book which should never have been torn up at all, an old book full of poetry.

'There's more of that lying about,' said the grocer. 'I gave an old woman some coffee-beans for it. If you'd like to give me sixpence, you can have the rest.'

'Thank you,' said the student. 'You can let me have it instead of the cheese. I can make do with bread and butter. It would be sinful to tear the whole book to pieces. You're a fine man, a practical man, but you understand no more about poetry than this cask here.'

Now that was a rude thing to say, especially as far as the cask was concerned, but the grocer laughed and the student laughed, for of course it was said as a kind of joke. But the goblin was annoyed that anyone dared speak like that to the grocer who was the landlord of the house and sold the best butter.

That night, when the shop was shut and everyone but the student was in bed, the goblin went in and took the wife's gift of the gab—she had no use for it while she was asleep—and whatever object in the room he put the gift of the gab on received the power of speech and could express its thoughts and feelings just as well as she could. But only one at a time could have it, which was a blessing, for otherwise they would all have talked at once.

And the goblin placed the gift of the gab on the cask where the old newspapers were kept. 'Is it really true,' he asked, 'that you don't know what poetry is?'

'Yes, of course I do,' said the cask. 'It's something you find at the bottom of the page in the newspaper and cut out. I should think I've more of it inside me than the student has, and I'm only a poor cask compared with the grocer.'

Then the goblin placed the gift of the gab on the coffee-mill—my, how it went! And he put it on the butter-tub and the till—they were all of the same opinion as the cask, and what most people are agreed upon you have to respect.

'Now I'll try it on the student!' And so the little goblin went quite quietly up the back stairs to the attic where the student lived. There was a light inside his room, and the goblin peeped through the key-hole and saw the student reading the tattered book from downstairs.

But how bright it was in the room! A bright beam of light rose out of the book and grew into a trunk, a mighty tree that rose up high and spread out its broad branches over the student. Every leaf looked so fresh, and every flower was the face of a lovely girl, some with eyes so dark and shining, others with eyes so blue and strangely bright. Every fruit was a shining star, and there was a sound of beautiful singing.

The little goblin had never even imagined anything so glorious, let alone seen and experienced it. And so he stayed there, standing tip-toe and peeping and peeping, until the light in the room went out. The student must have blown out his lamp and gone to bed, but the little goblin stayed where he was, for the song still sounded soft and lovely in his ears, a beautiful lullaby for the student who had lain down to rest.

'It's wonderful here,' said the little goblin. 'I hadn't expected anything like this.—I think I'll stay with the student—' And he thought about it—and thought about it very sensibly, and then he sighed and said, 'The student hasn't any porridge'—and so off he went—yes, he went down again to the grocer. And it was a good thing he did, for the cask had very nearly used up the good lady's gift of the gab telling everybody about everything that was inside it. It had already done this once from one angle, and now it was just about to turn over so that it could repeat it from the other angle, when the goblin came and took the gift of the gab back to the good lady. But from that time the whole shop, from the till to the firewood, based its views on what the cask said,

and they held it in such high esteem and had such confidence in it that ever after when the grocer was reading the art criticisms and theatre notices from the evening paper, they thought he had it all from the cask.

But the little goblin no longer sat quietly listening to all the wisdom and good sense down below. No, as soon as the light shone from the room in the attic, its beams were just like strong anchor-chains dragging him upstairs, and off he had to go and peep through the key-hole. Once there he was overpowered by a sense of greatness, such as we feel when we stand by the rolling sea and God passes over it in the stormy blast; and he would burst into tears—he did not know himself why he was crying—but there was something very blessed in those tears. How wonderful he thought it would be to sit under that tree with the student. But that could never be. And he was glad he had the keyhole.

He was still standing there in the cold passage when the winds of autumn were blowing down from the skylight. It was so very, very cold, but the little fellow did not feel it until the light went out in the attic and the sound of music died away in the wind. Ooh! Then he would freeze and creep downstairs again into his warm corner—how cosy and comfortable it was! And then his Christmas porridge came with its great lump of butter.—Yes, the grocer was the master for him.

But in the middle of the night the goblin was awakened by a frightful banging on the window-shutters. People outside were thumping on them; the watchman was blowing his whistle; a great fire had broken out and the whole street was lit up by the flames. Was it in this house or next door? Where was it? What a panic there was! The grocer's wife was so flurried that she took her gold ear-rings out of her ears and put them into her pocket so that she might at least save something. The grocer ran for his bonds, and the maid for her silk mantilla—for she was well enough off to have one. Everyone wanted to save what he prized most, and so, too, did the little goblin. In a couple of bounds he was up the stairs and in the attic with the student who was standing quite calmly by the open window and looking out at the fire which was in the house across the road. The little goblin seized the wonderful book from the table, put it into his red hood, and held it tight in both hands. The most precious treasure in the house had been rescued.

Then off he went, right out on to the roof, right up the chimney-stack, and there he sat lit up by the flames of the burning house opposite and holding closely in both hands his red hood and the treasure that lay inside it. Then he knew where his heart lay, whom he really belonged to.

But when the fire had been put out and he had become more level-headed, well—'I'll divide my time between them!' he said. 'I can't give up the grocer entirely, on account of the porridge.'

And that was really quite human! We, too, go to the grocer's—for porridge.

Five from a Peapod

THERE were once five peas in a peapod. They were green and the pod was green, and so they thought the whole world was green, and they were absolutely right. The pod grew and the peas grew. They adapted themselves to their accommodation and sat in a straight row. The sun shone outside and warmed the pod, the rain made it clean and clear; it was cosy and good inside, light in the daytime and dark at night, just as it should be. And as they sat there, the peas grew bigger and more and more thoughtful, for they had to do something.

'Must I always stay sitting here!' said one. 'I hope we don't grow hard with sitting so long. Isn't there something in store for us, something outside? I have a feeling there is.'

Weeks went by; the peas turned yellow, and the pod turned yellow. 'The whole world is turning yellow,' they said, and there was no reason why they shouldn't think so.

Then they felt a tug on the pod; it was pulled off, put into a human hand and then down into a jacket-pocket with several other well-filled peapods.

'We shall soon be opened now,' they said as they waited for it.

'I wish I knew which of us was going to travel farthest,' said the smallest pea. 'Well, we shall soon find out now.'

'Happen what may!' said the biggest.

'Crack!' The pod split open, and all five peas rolled out into the bright sunshine. They were lying in a child's hand. A little boy held them and said they were just the right peas for his pea-shooter. And with that one of the peas was put into his elder-twig pea-shooter and away it shot.

'I'm flying out into the wide world. Catch me if you can!' And it was gone.

'I', said the second, 'am going to fly straight into the sun; that will make a proper peapod, just the thing for me!'

Off it went.

'I'm going to sleep wherever I get to,' said the next two. 'But look, we're rolling off!' And they rolled on to the floor. But they too were picked up and put into the peashooter. 'We'll go farthest!'

'Happen what may!' said the last as it was shot into the air, and up it flew on to the old board under the attic window, right into a crack where there was moss and soft earth. The moss closed round it, and there it lay hidden, but not forgotten by Our Lord.

'Happen what may!' it said.

Inside, in the little attic, lived a poor woman who during the daytime went out polishing stoves, yes, and sawing firewood and doing heavy work, for she was strong and hard-working, but she was still as poor as ever. At home in her little room lay her only daughter, half grown-up, and very frail and delicate. A whole year had she lain abed, unable, it seemed, either to live or to die.

'She's going to her little sister!' said the woman. 'I had the two children, and it was hard enough for me to care for the two of them, but then the Lord went halves with me and took one to Himself. And now I do so want to keep the one I have left, but He'll not have them parted, and she'll go up to her little sister.'

But the sick girl stayed; she lay patiently and quietly all day long while her mother was out earning a bit of something for them.

It was spring-time, and early one morning just as mother was going out to work, the sun shone brightly through the little window and across the floor, and the sick girl looked up and over at the bottom pane of glass.

'What's that bit of green peeping up by the window-pane? It's moving in the wind.'

Her mother went over to the window and half opened it. 'Oh!' she said. 'Well! It's a little pea that's shot up with its green leaves. How did it get into the crack out here? There, now you've a little garden to look at.'

And the invalid's bed was moved nearer to the window where she could see the growing pea, and her mother went to work.

'Mother, I think I'm getting better!' the little girl said in the evening. 'The sun has been shining so warmly on me today. The little pea is doing well, and I want to do well, too, and get up and go out into the sunshine.'

'If only you could,' said her mother, but she did not believe it would happen. However, so that it should not break off in the wind, she stuck a little stick in beside the green shoot which had given her child the desire to live. She tied a piece of twine firmly to the board and to the top of the window-frame to give the shoot something to support itself with and to twist its tendrils round as it grew. And, in fact, it did grow, for you could mark its growth every day.

'My, just look, it's coming into flower!' said the woman one morning, and now she, too, began to hope and believe that her little sick girl would recover. It ran in her mind that the child's voice had grown more lively of late, that the last few mornings she had raised herself up in bed and looked with shining eyes at her little pea-garden with but one pea in it. The week after, the invalid was out of bed for the first time. She sat happily for over an hour in the warm sunshine with the window open, and outside a pale-pink pea blossom was in full bloom. The little girl bent her head down and very softly kissed its delicate petals. That day was a day of rejoicing.

'Our Lord Himself planted it and let it thrive to give you hope and joy, my dearest child, and to give me hope and joy, too!' said the happy mother, smiling at the flower as if it were a good angel from God.

But now the other peas! Well, the one that flew out into the wide world crying, 'Catch me if you can!' fell into the gutter under the roof and ended in a pigeon's crop where it lay like Jonah in the whale. The two lazy ones got just as far afield. They, too, were eaten by pigeons, and that, at any rate, served a useful purpose. But the fourth, who wanted to fly up to the sun, fell into the gutter in the street and lay there for days and weeks in the foul water, where it swelled up out of recognition.

'I'm getting beautifully fat!' said the pea. 'I shall split with fatness, and I should think there isn't a pea that can go farther than that, or ever has been! I'm the most remarkable of the five from the peapod!'

And the gutter agreed.

But up at the attic window stood the young girl with bright shining eyes and the flush of health upon her cheeks, and she folded her delicate hands over the pea blossom and thanked Our Lord for it.

'I'm keeping my own pea!' said the gutter down in the street.

Soup on a Sausage-Stick

or A Great Deal out of Nothing at All

I *Soup on a Sausage-Stick*

'WE had an excellent dinner yesterday!' said an old lady mouse to another mouse who had not been at the feast. 'I sat in the twenty-first place from the old Mouse King. That wasn't bad, was it? Now I really must tell you the menu. It was very well put together. First we had mouldy bread, then came bacon-rind and tallow-candle, and finally sausage—and then everyone had second helpings. It was as good as two dinners. There was a very cosy atmosphere with a lot of jolly nonsense, just like a family gathering. Nothing was left over, except the sausage-sticks. So we talked about them, and then someone brought up that old thing about making soup on a sausage-stick. Of course everyone had heard of it, but no one had ever tasted it, let alone been told how to make it. A toast was proposed, rather charmingly, to the inventor—"He deserved to be made Officer in charge of Poor Relief!" Wasn't that witty? And then the old Mouse King stood up and promised that whichever young female mouse could make the soup in the tastiest way should be his Queen. He gave them all a year and a day to find out how to make it.'

'That was reasonable enough,' said the other mouse. 'But how *do* you make that soup?'

Yes, how do you make it? That's what they were all asking, all the lady mice, young and old. They all wanted to be Queen, but they did not fancy the idea of going out into the wide world to find out the answer to the problem, and that would, of course, be necessary. It is not given to everybody to leave the family and the old cosy corners at home. Besides, you don't come across cheese-parings or smell bacon-rind every day out there in the wide world. Oh, no, you can find yourself starving, even, perhaps, being eaten alive by a cat!

Such thoughts were quite enough to frighten most of them from going out in search of knowledge. Only four mice presented themselves to undertake the journey. They were all young and lively, but poor. They were each to go to one of the four corners of the world, and so let fortune decide the winner. Each one took a sausage-stick with her to remind her why she was making the journey; it would also do as a pilgrim's staff.

They set out early in May, and early in May the year after they came back. But there were only three of them. The fourth did not return, nor had anything been heard of her, and now was the day when the decision was to be made.

'Some sorrow must always cloud our greatest joy,' said the Mouse King, but nevertheless he gave orders that all the mice from many miles around were to be invited to gather in the kitchen. The three mice who had returned from their travels stood in a row by themselves. A sausage-stick hung with black crape was set up for the fourth mouse who was missing. No one dared give an opinion before the three travellers had spoken and the Mouse King had added whatever else should be said.

Now we shall hear what happened!

II What the first little Mouse had Seen and Learnt on her Travels

'When I set out into the wide world,' said the first little mouse, 'I thought, as so many of my age do, that I'd absorbed all the wisdom of the world, but of course I hadn't; it takes a year and a day to do that. I went to sea at once and I took a ship that was sailing north. I had heard that when at sea a ship's cook knows how to look after himself, but it's easy enough to look after yourself when everywhere is filled with sides of bacon, barrels of salt meat and flour full of mites. You live in luxury, but you don't learn anything about making soup on a sausage-stick. We sailed for many days and nights, rolling and pitching and getting wet. When we arrived where we were bound for, I left the ship; that was far up in the north.

'It's strange leaving your own cosy corner at home, boarding a ship, which is a kind of little corner, too, and then suddenly finding yourself over a hundred miles away in a foreign country. There were trackless forests of fir and birch, their scent so powerful that I didn't care for it at all. The wild plants, too, had such a spicy smell that they made me sneeze and think of sausage. There were great lakes in the forest, their water quite clear when you were near to them, but as black as ink when you saw them from a distance. There were white swans lying so still on the water that I took them for foam, but when I saw them flying and walking, I knew what they were. They belong to the goose family, as you can tell by the way they walk. You can't hide your family traits!

'I kept to my own kind, and joined the wood-mice and the field-mice, who, incidentally, knew precious little, especially about entertaining and preparing food, and that was just what I was travelling abroad to find out about. The very idea of making soup on a sausage-stick seemed to them such an extraordinary notion that the news of it spread like wild fire through the whole forest. But to find an answer to the problem they reckoned utterly impossible. So least of all did I think that there, and that very same night, I should be initiated into the making of it!

'It was midsummer, and that, they said, was why the forest was so strongly scented, the plants so spicy, and the lakes so clear and yet so dark with the white swans floating on them. On the edge of the forest, among three or four houses, a pole had been set up, as tall as a mainmast, and from the top hung garlands and ribbons. It was a Maypole, with girls and lads dancing round it, singing and vying with the fiddler's music. It went on merrily until sunset and on into the moonlight, but I didn't join in. What can a little mouse do at a forest dance? I sat in the soft moss and held on to my sausage-stick. The moon was shining particularly brightly on one spot where there was a tree covered in moss as fine—yes, I dare say it—as fine as the Mouse King's fur, but it was green in colour and so it was good for the eyes. Then all at once there came marching along the most charming little people, no taller than my knee. They looked like humans, but they were better proportioned. They called themselves elves, and they wore fine clothes of flower petals inset with the wings of flies and midges—not at all bad looking!

'No sooner had they arrived than they seemed to be searching for something, I didn't know what, but then two or three came over to me, and the most important among them pointed to my sausage-stick and said, "That's just what we want! It's cut out for the very purpose. It's splendid." And he grew more and more delighted as he looked at my pilgrim's staff.

'"Certainly borrow, but not keep!" I said.

'"Not keep!" they all said, taking hold of the sausage-stick which I let go, and dancing with it over to the plot of fine moss. They set it up there in the middle of the green. They wanted a Maypole, too, and the one they now had might have been cut purposely for them. They soon decorated it, and a splendid sight it was!

'Little spiders spun gold thread round it, and hung up waving veils and banners, so delicately woven, bleached so snowy white in the moonshine that my eyes were dazzled. They took colours from butterfly-wings and sprinkled them over the fine white lawn, and flowers and diamonds shone upon it. I no longer recognized my sausage-stick. The like of such a Maypole as they had made of it was surely not to be found in the whole wide world. Then the main party of elves arrived. They wore no clothes at all, and nothing could be more exquisitely formed than they were. I was invited to watch their revelry, but from a distance, for I was too big for them.

'Now the music began! It sounded as if thousands of glass bells were ringing, full and strong. I thought there were swans singing, yes, I thought, too, I could hear the cuckoo and the thrush, and at last it seemed as if the whole forest was joining in. There were children's voices, the chiming of bells and birdsong, the loveliest of melodies, and all that delightful music rang out from the elves' maypole, and that was my sausage-stick! I would never have believed that so much could come out of it, but it depends, of course, upon whose hands it comes into. I was really so moved that I cried for pure joy like a little mouse.

'The night was all too short. But the nights are not very long up there at that time of year. With the break of day there came a fresh breeze. The mirror-like surface of the forest lake was ruffled, the fluttering veils and banners flew away into the air, the swaying fretwork of cobweb—hanging bridges, balustrades, whatever you may call them—that was spun from leaf to leaf dissolved into nothing. Six elves brought me back my sausage-stick, asking at the same time if I had a wish they could grant. And so I asked them to tell me how to make soup on a sausage-stick.

'"How do we set about it?" said the leader, and he laughed. "But you've only just seen how we do it! You scarcely recognized your own sausage-stick, did you? We made soup on it—something out of nothing!"

'"You mean, that's it?" I said, and told them straight out why I was on my travels and what was expected of me at home. "What good," I asked, "will it do the Mouse King and all our vast kingdom that I've seen these beautiful things. I can't shake them out of the sausage-stick and say, 'Look, here's the stick, and now comes the soup!' It would, I suppose, always make a sort of second course after they'd eaten their fill!"

'Then the elf dipped his little finger into a blue violet and said to me, "Now pay attention; I'm smearing your staff, and when you get home to the Mouse King's palace, you must touch your king's warm breast with it, and then violets will spring out all over it, even in the coldest of the winter time. There, now you have something to take home with you, and I'll give you a little bit more as well!"'

But before the little mouse told what this little bit was, she turned her staff towards the King's breast, and there really did spring from it the most beautiful bouquet of flowers! Its perfume was so overpowering that the Mouse King ordered the mice who were standing nearest to the fireplace to stick their tails into the fire so that they could get a little smell of singeing. The scent of the violets was not to be borne; it was not the kind of smell they liked at all!

'But what was that little bit more you spoke of?' asked the Mouse King.

'Oh, yes,' said the little mouse, 'it's what they call an effect, you know.' And then she turned the sausage-stick round, and there were no longer any flowers on it. She was holding nothing but a bare stick, which she now raised like a conductor's baton.

'"Violets are for sight, smell and touch," the elf told me, "but here is something else for hearing and taste!"'

And so the little mouse began beating time. Music sounded, but not like that that had rung out in the forest at the elves' revelry; oh, no, this was the kind that can be heard in the kitchen. What a racket! It came without warning, with a noise like the wind whistling and howling through all the chimneys; kettles and pots boiled over, and the coal-shovel thundered on the brass cauldron. Then, just as suddenly, it was quiet again and you could hear the tea-kettle's muffled singing—such a strange song, you couldn't tell whether it was beginning or ending. And then the little pot boiled and the big pot boiled, neither taking the slightest notice of the other, as if they were giving no thought at all to what they were doing. And the little mouse swung her baton more and more wildly—the pots seethed, bubbled and boiled over, the wind howled, the chimney whistled. Whew! it was so alarming that the little mouse herself lost control of her stick and dropped it.

'A very tricky soup!' said the old Mouse King. 'Is it ready now?'

'That's all there is,' said the little mouse, curtseying.

'That all! Well, let's hear what the next one has to say,' said the Mouse King.

III *What the Second Little Mouse had to Tell*

'I was born in the palace library,' said the second mouse. 'Neither my family nor I have ever known the good fortune of getting into the dining-room, let alone the larder. I first saw a kitchen when I went on my travels, and now here today, of course. We went really hungry quite often in the library, but we acquired a great deal of knowledge. The rumour reached us up there of the royal prize offered for making soup on a sausage-stick. Then it was that my grandmother pulled out a manuscript—she couldn't read it, but she had heard it read—in which it said, "If you are a poet, you can make soup on a sausage-stick." She asked me if I were a poet. Not as far as I knew, I said, and she said that I'd better go and see about becoming one.

'"But what's required for it?" I asked, for it seemed to me just as difficult to find out how to be a poet as to make the soup.

'But grandmother had heard people reading and she said that three things were chiefly necessary. "Intelligence, fantasy and feeling!" she said. "If you can go and get them inside you, then you'll be a poet, and you'll get on all right with the sausage-stick."

'And so I went westwards out into the wide world to become a poet.

'Intelligence, I knew, was most important in everything—the other two are not held in the same estimation. So I went out for intelligence first. Yes, but where do you find it? Go to the ant and learn wisdom, a great king of the Land of Israel once said. I knew that from the library, and so I didn't stop until I came to the first big ant-hill, and there I lay in wait to become wise.

'They are a very respectable set of creatures, the ants. They are intelligence itself. Everything about them is like a sum that has been done correctly and works out right. Working and laying eggs, they say, is living in the present and looking after the future, and so that's what they do. They are divided into clean ants and dirty ants, and their rank is denoted by a number.

'The Queen Ant is Number One, and her opinion is the only correct one. She has absorbed all their wisdom—and it was important for me to know that. She said a great deal that was so clever that I thought it stupid. For example, she said that their ant-hill was the tallest thing in the world, yet close by stood a tree that was taller, much taller. That couldn't be denied, and so the ants didn't talk about it. One evening an ant lost its way near the tree and crawled up the trunk, not even as far as the crown of the tree, but nevertheless higher than any ant had been before, and then it turned back and found its way home. Back in the ant-hill the ant told the others that there was something much taller outside. All the other ants considered this an insult to the whole community, and so the ant was condemned to wear a muzzle and sent into solitary confinement for life. Shortly afterwards, however, another ant came to the tree, and made the same journey and the same discovery. It, too, spoke about it, but in a cautious and vague way, and as it was a respected ant, one of the clean ones, it was believed and when it died, they erected an egg-shell as a memorial to it, for they esteem all branches of knowledge.

'I saw,' said the little mouse, 'that the ants continually ran about with their eggs on their backs. One of them dropped hers; she made a great effort to get it up again, but without success. Then two others came along and helped with all their strength, until they were in danger of losing their own eggs. At that point they stopped at once, for you must consider yourself first. The Queen Ant said of this incident that they had shown both feeling and intelligence. "Those two qualities," she said, "place us ants highest among rational beings. Intelligence must and should be the first consideration, and I have the greatest amount of it!" With that she stood up on her hind legs, and was so easy to recognize that I could not mistake her. And so I swallowed her! "Go to the ant and learn wisdom." Now I'd got the Queen herself!

'I now went over to the great tree I told you of. It was an oak with a tall trunk and huge spreading branches, and was

245

very old. I knew that a living being dwelt inside it, a woman—a dryad, she's called—who is born with the tree and dies with it. I had heard about this in the library, and now I actually saw such a tree with its oak-maiden. She gave a dreadful shriek when she saw me so near. Like all women she was scared of mice, but she had more reason to be, for I could gnaw through the tree, and her life depended upon it. I spoke to her in a straightforward friendly way, and that gave her courage. She picked me up and held me in her delicate hand, and when she found out why I had gone out into the wide world, she promised that perhaps that very evening I should get one of the two treasures I was still looking for. She told me that Fantasy was her very good friend, that he was as beautiful as the God of Love, and that he would often come and rest under the leafy branches of her tree. Then it would rustle over their heads all the more strongly because he was there, too. He called her his own dryad, she said, and the tree his own tree.

'The magnificent gnarled oak-tree was after his own heart, with its roots spreading deeply and firmly into the ground, its trunk and crown rising high into the fresh air; aware of the driving snow, the keen winds and the warm sunshine, as one should be aware of them.

'"Yes," the dryad said, "the birds sing up there and tell of foreign lands! And on that one dead bough the stork has built a nest; it looks well, and we hear a little from the land of the pyramids. Fantasy enjoys all that, but it is not enough for him, and so I have to tell him about my own life in the forest from the time when I was little and the tree so small that a nettle could hide it, until now when it has grown so great and strong. Now you sit under the woodruff and keep your eyes open. When Fantasy comes, I shall no doubt find an opportunity to give him a little nip on one of his wings and pull a small feather out. Take it; no poet ever had anything better. Then you will have all you need."

'Fantasy came, the feather was plucked out, and I seized

it,' said the little mouse. 'I kept it in water until it was soft. It was still very difficult to digest, but I managed to chew it up. It's not at all easy to chew your way into becoming a poet, there's so much you have to get inside you.

'I now had two things, intelligence and fantasy, and with their help I now knew that the third was to be found in the library. A great man has both said and written that there are novels whose only merit is to relieve people of their superfluous tears, being, in fact, a kind of sponge to soak up their feelings. I remembered two or three books of this sort; they had always looked very appetizing to me. They had been so much read, and were so well-thumbed and greasy, that they must have absorbed no end of powerful feeling.

'I went home to the library and immediately ate through practically a whole novel, the soft part, that's to say, and that's the important bit. The rind, the binding as they call it, I left alone. By the time I'd digested this book and another one as well, I already felt something working inside me. I ate a bit of the third one, and then I was a poet. Anyway, that's what I told myself, and I told the others so, too. I had a headache and a tummy-ache, and I don't know what other aches besides. I now thought over what stories I could tell that might have anything to do with sausage-sticks, and my thoughts were so filled with sticks of all kinds that only then did I realize what a remarkable intelligence the Queen Ant I'd swallowed must have had. I remembered the man who put a white stick in his mouth, and both he and the stick became invisible. I thought of a pin in a tankard of old ale, of standing on your pins, putting a spoke in one's wheel or a nail in one's coffin. All my thoughts ran on sticks and pins and pegs, and poems could be written about them if one were a poet; and that's what I am—I've worn myself out to become one. And so every day of the week I shall be able to serve you a stick—I mean a story, and that's my soup!'

'Let's hear the third mouse then,' said the Mouse King.

'Peep, peep!' came a squeak from the kitchen door, and a

little mouse, the fourth one, the one they had thought was dead, scurried in, knocking down the sausage-stick with the black crape on it. She had been running night and day, she had travelled on the railway by goods-train whenever she had the chance, and yet she was only just in time. She pushed her way forward quite out of breath. She had lost her sausage-stick but not her voice, and she began speaking right away as if they were only waiting for her and only wanted to listen to her, as if nothing else in the world mattered to anyone. She began talking at once, and she did not stop until she had said all she had to say. She had arrived so unexpectedly that no one had time to pass remarks about her or comment on what she was saying all the time she was talking. And now we must listen.

IV   What the Fourth Mouse, who Spoke before the Third Mouse had Spoken, had to Tell

'I went straight to the biggest city I could find,' she said. 'I forget its name—I'm not very good at names. From the railway station I went with some goods that had been confiscated to the police court in the Town Hall, and there I made friends with the gaoler. He told me about his prisoners, especially one who had made reckless speeches about which there had been no end of talking and reading and writing. "The whole thing," the gaoler said, "is soup on a sausage-stick! But that soup may cost him his head."

'And that aroused my interest in the prisoner,' said the little mouse, 'and I watched for an opportunity to slip into his cell—there's always a hole for a mouse to get through behind locked doors. He was a pale-looking man with a big beard and large bright eyes. His lamp was smoking, but the walls were used to that, and it made them no blacker than they already were. The prisoner had scratched pictures and verses on them, in white on black, but I didn't read them. I think he was bored; anyway, I was a welcome guest. He tempted me with breadcrumbs and enticed me with whistling and gentle words. He was delighted to find me there; I trusted him, and we became friends. He shared his bread and water with me, and gave me cheese and sausage; I lived well, but it was chiefly our good companionship that kept me there. He would let me run up his hand and arm, right up his sleeve, and crawl into his beard. He called me his little friend; I grew really fond of him—an attachment like that is bound to be mutual. I forgot about my business out in the wide world. I lost my sausage-stick in a crack in the floor—it's still there somewhere.

'I wanted to stay where I was. If I went away, then the poor prisoner would have no one at all, and that's all too little in this world. So I stayed—but he didn't! He talked to me quite sadly the last time we were together, gave me twice as much bread and cheese-rind as he usually did, and then kissed his hand to me. He went away and never came back. I don't know his story.

'"Soup on a sausage-stick!" said the gaoler. I went to him, but I shouldn't have trusted him. He took me on his hand, but he put me in a cage, in a treadmill. That's awful! You run and run and get no farther, and they only laugh at you.

'The gaoler's grand-daughter was a pretty little thing with curly golden-yellow hair, merry eyes and a laughing mouth. "Poor little mouse," she said, peeping into my dreadful cage. She pulled out the iron pin, and I jumped down on to the window-sill and out into the gutter. Free, free!—That was all I cared about. I gave no thought to the purpose of my travels.

'It was dark and getting on for night-time, and so I took refuge in an old tower. A watchman and an owl lived there; I trusted neither of them, least of all the owl. Owls look like cats and have one great fault—they eat mice! But one can be mistaken, and this time I was. She was a very respectable, highly cultured old owl. She knew more than the watchman, and just as much as I did myself. The young owlets made a great to-do over everything. "Don't make soup on a sausage-stick," she would say, and that was as severe as she could bring herself to be with them, so great was her affection for her family. I felt so much confidence in her that I said, "Peep!" from the crack where I was sitting. This mark of my trust pleased her, and she assured me that she would take me under her protection. She would allow no creature to harm me—that she would do herself when winter came and they were short of food.

'She was clever at everything. She pointed out to me that the watchman could not hoot unless he used a horn which he carried loosely slung over his shoulder. "He fancies himself dreadfully with it—he thinks he's the Owl in the Tower. All bark and no bite. Soup on a sausage-stick!" I asked her for the recipe, and so she explained it to me, "Soup on a sausage-stick is only a manner of speaking that men use. They give various meanings to it, and everyone thinks his own is the right one; but it really amounts to nothing at all."

'"Nothing!" I said. I was stunned. The truth isn't always pleasant, but it's the highest and best we know. The old owl said that, too. I thought it over, and it struck me that if I brought back the best, then I should bring back much more than soup on a sausage-stick. And so off I hurried to get home in good time, and bring the highest and best—the truth. We mice are enlightened creatures, and the Mouse King is above us all. He could make me his Queen for the sake of the truth.'

'Your truth is a lie!' said the mouse who had not yet had a chance to speak. 'I can make the soup, and I'm going to do it!'

V *How it was Made*

'I haven't been travelling,' said the fourth mouse. 'I stayed in my own country, and that was the right thing to do. You don't need to travel; you can get everything just as well here. So here I stayed. I haven't learnt the answer from super-natural beings, I haven't eaten myself into it, nor have I talked with owls. I got it through my own independent thinking. So will you just put the kettle on, full of water, right up to the top. Light the fire under it. Let it burn till the water boils—it must boil right over. Throw the sausage-stick in! Now, will the Mouse King be pleased to stick his tail right into the boiling water and give it a stir? The longer he stirs, the stronger the soup will be—and it costs nothing. There's no need for any seasoning—just keep stirring.'

'Can't somebody else do it?' asked the Mouse King.

'No!' said the mouse. 'The power lies only in the Mouse King's tail.'

The water boiled over, and the Mouse King came and stood close to it—it was pretty dangerous!—and he stuck his tail out as mice do in the dairy when they skim the cream off a pan of milk and then lick their tails. But he got no farther than the hot steam, and then jumped straight down again.

'Of course you shall be my Queen!' he said. 'But we'll wait for the soup until our golden wedding. Then the poor of my kingdom will have something to look forward to, and for quite a long time!'

And so they were married. But when they got home after the wedding, several mice said, 'You can't really call it soup on a sausage-stick; it was more like soup on a mouse-tail.' One or two things in the stories they had heard they found quite good, but they thought the composition as a whole might well have been different. 'Now I would have told it like this...!'

That was criticism, and that's always so clever—afterwards!

The story went right round the world. Opinions about it were divided, but the story itself remained intact, and that is as it should be, in big things and little things, and in soup on a sausage-stick. Only you mustn't expect to be thanked for it.

Absolutely True!

'IT'S a shocking story!' said a hen, who lived in a part of the town far from where the incident had taken place. 'There's a shocking story being told about a hen-house. I daren't sleep alone tonight! It's a good thing there are lots of us together on the perch!'

And so she told her tale, and the other hens' feathers stood up on end and the cock's comb drooped down. It's absolutely true!

But we'll begin at the beginning, and that took place in a hen-house on the other side of the town. The sun went down and the hens flew up to roost. One of them was white-feathered, short in the leg, laid her regulation egg and was, for a hen, respectable in every way. Well, as she flew on to the perch, she pecked herself with her beak and a little feather flew out.

'There it goes,' she said. 'The more I pluck, the prettier I shall be.' Now this was said in fun, for she was the life and soul of the hen-house, but otherwise, as we have said, very respectable. And so she fell asleep.

It was dark all round, with one hen sitting close up to another, and the hen who sat nearest the one who had lost a feather was not yet asleep. She heard and she did not hear, as you have to in this world in order to preserve your peace of mind. But nevertheless she could not resist saying to her other neighbour, 'Did you hear what was said just now? I name no names, but there's a hen who wants to pluck herself to look attractive. If I were the cock, I'd treat her with scorn!'

Now immediately above the hen-house sat Mrs. Owl with Mr. Owl and the little owls. They have sharp ears in that family, and so they heard every word that the second hen said. They rolled their eyes, and mother-owl fanned herself with her wings. 'Just don't listen! But I suppose you heard what was said? I heard it with my own ears, and you have to hear a great deal before they fall off. There's one of the hens who has so far forgotten herself that she's sitting and plucking all her feathers out, and letting the cock watch her.'

'Prenez garde aux enfants!' said father owl. 'It's not a fit subject for children's ears!'

'But I must just tell the owl over the way. She's such a respectable creature.' And off flew mother.

'Whoo-hoo! Whoo-hoo!' they both hooted to the pigeons in the dove-cot across the way. 'Have you heard? Have you heard? Tu-whoo! There's a hen that's plucked all her feathers out, and all on account of the cock. She's freezing to death—if she isn't dead already, whoo-hoo!'

'Where? Where?' cooed the pigeons.

'In the yard across the way! I as good as saw it myself. It's hardly a nice story to tell, but it's absolutely true!'

'I believe you, I believe you, every single word!' said the

pigeons, and they cooed down into their own poultry-yard, 'There's a hen—some say two—who's plucked all her feathers out to make herself look different from the rest to attract the cock's notice! It's a risky game to play—you could catch cold and die of a fever. And she's dead now—they both are!'

'Wake up! Wake up!' crowed the cock, flying up on to the fence. His eyes were still heavy with sleep, but he crowed all the same. 'There are three hens who've died of hopeless love for a cock. They plucked all their feathers out! It's a grim story, but I don't like to keep it to myself. Pass it on!'

'Pass it on!' squeaked the bats, and the hens clucked and the cocks crowed, 'Pass it on! Pass it on!' And so the story travelled from hen-house to hen-house, and at last came back to the place where it had started.

'There are five hens,' it went, 'who've all plucked their feathers out to see which of them had lost most weight from being love-sick for the cock. Then they pecked one another until they drew blood and fell down dead, to the shame and scandal of their family and the loss of their owner.'

And the hen who had lost her loose little feather naturally failed to recognise her own story; and as she was a respectable hen, she said, 'I despise such creatures! But there are too many of that sort! A story like that shouldn't be hushed up, and I shall do my best to see that it gets into the papers! Then it will go all over the country. That's what hens like that deserve, and their families, too.'

And so it was sent to the paper and printed, and what is absolutely true is that one little feather can turn into five hens.

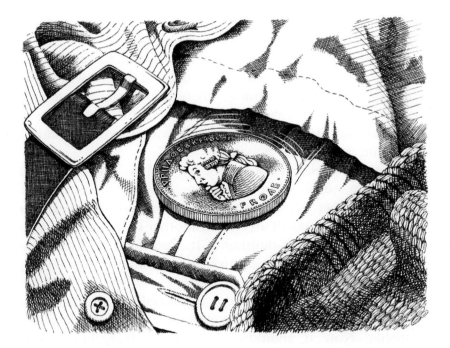

The Silver Shilling

THERE was once a shilling that came bright and shining from the mint, dancing and ringing out, 'Hurrah! Now I shall go out into the wide world.' And so it did.

The child would hold it fast in its little hot hands, and the miser in his cold clammy ones. The elderly would turn it round and turn it over many times, while the young would send it spinning on its way as soon as they got it. The shilling was made of silver, with a very little copper in it, and had already been a whole year out in the world, that is to say, out and about in the country where it was minted. Then it found itself travelling abroad. It was the last of the country's coinage left in the purse the traveller had taken with him, and he did not know he had it until he found it between his fingers.

'Well, I've still got a shilling from home,' he said. 'It can travel with me.' And the shilling rang and danced with joy as he put it back into his purse. There it lay among its foreign companions that came and went, one making room for another, but the shilling from home was always left behind; and that was a great honour.

Several weeks had now gone by, and the shilling was far out in the world, without knowing exactly where. It heard from the other coins that they were French or Italian, one said they were now in this town, another in that one. But the shilling could make nothing of it. You don't really see the world when you are in a purse all the time as the shilling was.

Then one day as it lay there, it noticed that the purse was not shut, and so it slipped towards the opening to take a look outside. Now it shouldn't have done that, but it was inquisitive, and that doesn't pay. It slid out into the trouser-pocket, and when the purse was taken out that evening, the shilling was left behind, and there it stayed when the clothes were put out in the corridor to be pressed. As they were dropped on the floor, it fell out; no one heard it, no one saw it.

Early in the morning the clothes were brought in, the traveller put them on, and off he went—but without the shilling. It was found and taken out with three other coins to be used once more.

'It's fine to look round the world,' thought the shilling, 'to get to know different peoples and different customs.'

'What sort of a coin's that?' someone said at that moment. 'That's no coin of the realm. That's false, that is! No good at all!'

And that's where its story begins, as it was told afterwards by the shilling itself.

'False! No good! The words went right through me,' said

the shilling. 'I knew I was made of good silver, with a good ring and a proper stamp. They must have made a mistake; it couldn't be me they meant. But it was me all right! It was me they called false! Me that was no good! "I must spend it after dark!" said the man who had picked me up. That's what he did, and in the daylight I was once again given a scolding: "False! No good! We must see about getting rid of it."

'"Wretched shilling that I am," I thought, "what good to me are my silver, my value, my stamp, when they mean nothing! In this world you're worth no more than what the world is prepared to believe of you. It must be dreadful to have a bad conscience and slink along evil ways, when I, who am completely innocent, feel like this simply on account of my appearance."

'Every time I was taken out, I dreaded the eyes that would look at me; I knew I should be pushed back, thrown on to the counter like a liar and a swindler.

'Once I came into the hands of a poor old woman, who had received me as part of her day's wages for her toil and hard work, and she couldn't get rid of me at all. No one would take me; I was a bit of real bad luck for her.

'"I'm forced to cheat someone with it, and that's the truth," she said. "I can't afford to keep a false shilling. The baker shall have it; he can best stand the loss, he's rich enough; but it's wrong I'm doing for all that."

'"Must I now be a burden on this poor old woman's conscience?" I sighed. "Have I really changed so much in my later days?"

'The woman went to the wealthy baker, but he knew all too well what coins were legal currency. I was thrown back into the woman's face, and she got no bread for me. I felt sad at heart to think that I had been minted to harm others like that—I, who in my younger days had been so confident, so sure of myself, so certain of my value and the genuineness of my stamp. I grew as melancholy as a poor shilling can be when no one will have it. But the woman took me back home, regarded me thoughtfully in a kind and friendly way, and said, "No, I'll not try to fool anyone with you. I'll make a hole in you so that everyone can see you're a false thing. And yet ... I've an idea now ... perhaps you're a lucky shilling; yes, I believe you are! A thought's come into my head. I'll bore a hole in the shilling, thread a lace through the hole and give it to the neighbour's little girl to wear round her neck as a lucky coin."

'And so she pierced a hole in me; it's never comfortable to have a hole pierced in you, but when the cause is good, you can stand a lot. I had a lace put through me and became a sort of medal. I was hung round the little girl's neck, and she smiled at me and kissed me, and I rested all night on the child's warm innocent breast.

'Early in the morning her mother took me between her fingers, looked at me, and had ideas of her own about me, as I quickly realized. She took out a pair of scissors and snipped through the lace.

'"A lucky coin," she said. "Well, now we shall see!" She laid me in acid that turned me green, then she puttied the hole up, polished me up a little, and then when it was growing dark, she went out with me to buy a lottery ticket that was to bring her good luck.

'How uneasy I felt! I had such a tight feeling, I thought I should snap in two; I knew I should be called false and thrown out, and right in front of all those other coins, too, that lay there with inscriptions and heads they could be proud of. But I slipped through; there were so many people buying tickets and the collector was so busy that I went tinkling into the drawer amongst all the other coins. Whether the ticket won anything I don't know, but I do know that the very next day I was picked out as a false coin, laid aside, and sent out to cheat again, always to cheat! And that's something you can't stand when you have an honest character, and that I can't deny I have.

'For long enough I was passed like this from hand to

hand, from house to house, always sworn at, always frowned upon. No one had faith in me, and I had no faith in myself, or in the world either; it was a difficult time.

'Then one day there came a traveller, who was, of course, imposed upon. He was unsuspecting enough to take me for current coin, but when he was going to spend me, I again heard the cry, "No good! False!"

'"But it was given to me as genuine," said the man, looking at me closely. Then he smiled all over his face, and that was something I wasn't used to when people took a close look at me. "Well, what about that!" he said. "It's a coin from our own country, a good honest shilling from home, that someone's made a hole in and called false! That's an odd thing, that is! I'll keep you and take you home with me!"

'A flood of happiness swept over me. I'd been called a good honest shilling, and I was going home where one and all would recognize me and know I was made of good silver and had a genuine impression. I could have thrown out sparks of joy, but it's not in my nature to spark; steel can, but not silver.

'I was wrapped in fine white paper so that I should not be muddled with the other coins and get lost. Only on festive occasions when people from my own country met together, was I taken out and shown round and highly spoken of. They said I was interesting. It's very odd that one can be interesting without saying a single word!

'And so I came home again. My troubles were all over, and my happiness began anew, for I was of good silver, I had a genuine stamp, and the fact that I had a hole made in me to mark me as false did me no harm at all—that doesn't matter, as long as one isn't false. You just have to hold out, and all will come right in time. That's my belief,' said the shilling.